ANALYTICAL METHODS OF OPTIMIZATION

ANALYTICAL METHODS OF OPTIMIZATION

D. F. LAWDEN
Emeritus Professor
University of Aston in Birmingham, U.K.

DOVER PUBLICATIONS, INC.
Mineola, New York

Copyright

Copyright © 1975, 2003 by D. F. Lawden
All rights reserved.

Bibliographical Note

This Dover edition, first published in 2006, is an unabridged republication of the work originally published in 1975 by the Scottish Academic Press, Ltd., Edinburgh and London.

Readers of this book who would like to receive the solutions to the exercises may request them from the publisher at the following e-mail address: editors@doverpublications.com

International Standard Book Number: 0-486-45034-1

Manufactured in the United States of America
Dover Publications, Inc., 31 East 2nd Street, Mineola, N.Y. 11501

PREFACE

My main object in writing this book has been to present the classical theory of the calculus of variations in a form which is most appropriate for application to modern problems of optimizing the behaviour of engineering systems. Two areas in which such problems arise are in the design of control systems and in the choice of rocket trajectories to be followed by terrestrial and extra-terrestrial vehicles. I have had experience of research in both these fields and this will undoubtedly have influenced the manner in which the material has been presented. However, it is hoped that the book provides a sufficiently elaborate and general introduction to the basic ideas of optimization theory to enable the reader to follow current applications by applied mathematicians and theoretical engineers to a very wide diversity of problems.

It is intended that the book should be as helpful as possible to applied mathematicians who are primarily concerned to solve practical problems and who therefore wish to become acquainted with the relevant techniques as expeditiously as possible without any unnecessary digression into pure mathematical minutiae. Thus, although an attempt has been made to maintain the argument at a high level of logical rigour, generality of statement has invariably been sacrificed where it was felt that this would accelerate the reader's understanding without significantly reducing the field of practical applicability of the results. This emphasis on method and principle is also indicated by the large number of illustrative worked problems which will be found scattered throughout the text and the sets of exercises for working by the reader which have been collected at the end of each chapter.

As the title of the book makes clear, numerical optimization techniques are not included within its scope. Although it has to be admitted that the great majority of practical problems in this field can only be given a complete solution by the use of these techniques, it is impossible to arrive at a full understanding of numerical methods without making a study of their analytical basis. Further, by applying the analytical methods to a first approximate mathematical model of a given physical system, a good idea of the nature of its optimal behaviour can often be obtained; this will be of great value in directing the course of the subsequent numerical calculation of a more exact solution.

It is assumed that the reader has completed a course in the analysis of functions of several real variables and is familiar with the elementary theory of ordinary differential equations, especially linear equations. Such a course is usually completed by the end of the second year of study for a university degree in mathematics and the book is accordingly suitable for study by applied mathematicians in their final year. It will also be appropriate for use as a text by post-graduate M.Sc. or diploma students making a special study of systems' theory.

The repeated index summation convention has been employed throughout the book. Unless otherwise stated, it may be assumed that the indices i, j, k, range over the integers from 1 to M and the indices r, s, t range from 1 to N; other indices which are used less frequently, together with their ranges, are as follows: $l(1$ to $P)$, $m(1$ to $Q)$, $n(1$ to $Q + 1)$, $p(1$ to $S)$, $q(1$ to $T)$. Initial values of variables are distinguished by a superscript 0 and final values by a superscript 1. The end of every worked problem is indicated by a dot thus, ●.

The suggestion that I should write this book came from Professor Alan Jeffrey and I should like to take this opportunity of thanking him for his kind encouragement. Quite as arduous as writing the book was the task of typing it and for this service and for emerging cheerful but perhaps not unscathed from the thickets of superscripts and subscripts, I have to thank my secretary Mrs. Audrey Breakspear.

D. F. LAWDEN

Mathematics Department
University of Aston in Birmingham.
January, 1974

CONTENTS

Preface   v

CHAPTER 1   STATIC SYSTEMS   1
1.1   Statement of the Problem   1
1.2   Unconstrained Control   2
1.3   Equality Constraints   7
1.4   The Lagrange Multipliers   13
1.5   Inequality Constraints   14
1.6   Linear Programming   17

CHAPTER 2   CONTROL SYSTEMS   24
2.1   Definitions   24
2.2   Linear Control Systems   26
2.3   Infinite Series of Matrices   30
2.4   Autonomous Linear Systems   33
2.5   Optimal Control   36
2.6   Necessary Conditions for Optimal Control   37
2.7   Optimization of Linear Systems   47
2.8   Escape from a Circular Orbit   53

CHAPTER 3   ADDITIONAL CONSTRAINTS   62
3.1   Constraints on the Final State   62
3.2   An Integral of Hamilton's Equations   72
3.3   Calculus of Variations Problem   78
3.4   Non-differential Constraints   81
3.5   Integral Constraints   86
3.6   Discontinuous Controls   88
3.7   Linear Systems   94

CHAPTER 4   HAMILTON-JACOBI EQUATION   103
4.1   Optimal Cost Function   103
4.2   Hamilton-Jacobi Equation   104
4.3   Derivatives of the Optimal Cost Function   108
4.4   Pontryagin's Principle   111
4.5   Optimal Rocket Trajectories   116
4.6   Weierstrass and Clebsch Conditions   120

CONTENTS

CHAPTER 5 THE ACCESSORY OPTIMIZATION PROBLEM 123
- 5.1 Second Variation of the Cost 123
- 5.2 Accessory Problem. Conjugate Points 126
- 5.3 Solution of the Accessory Equations 129
- 5.4 The Accessory Riccati Equation 133
- 5.5 Sufficiency Conditions 138

Bibliography 153
Index 155

1 STATIC SYSTEMS

1.1 Statement of the problem

The fundamental problem which we shall study can be expressed in the following terms: Suppose the state of a given physical system can be changed by varying the values of N physical quantities denoted by $u_1, u_2, \ldots, u_N$. This process whereby the system's state is influenced by an external agent or operator will be referred to as *controlling the system* and the quantities u_r $(r = 1, 2, \ldots, N)$ will be called the *control variables*. The behaviour of the system or the configuration it adopts as a consequence of any particular choice of control variables, we shall designate the system's *response*. It will be desired to control the system in such a manner that its response is as close as possible to a standard the operator has his own reasons for striving to attain. It will always be assumed that the operator's success in achieving this object can be measured by giving the value of a quantity C called the *performance criterion* (or *index*). Thus, a response resulting in a large value of C may be judged superior to any other response which gives a smaller value. In these circumstances, our problem is to choose values for the control variables which maximize C. The system response which results from such a choice will be said to be *optimal* and our problem may accordingly be described as the *optimization of systems*.

In some cases, it is more natural to assess the performance of a system by reference to the cost to the operator of eliciting a particular response. The quantity C will then be termed the *cost* or *penalty index* and the optimization problem will be to minimize its value.

In this chapter, we shall investigate the optimization of systems whose responses to control are of a particularly simple character. Thus, having set the control variables to chosen values, it will be assumed that the system under consideration responds by adopting a corresponding fixed configuration and that the excellence or otherwise of this configuration for the operator's purpose can be assessed by calculation of an index C; i.e. C will be a function of the control variables. For example, if the system is a yacht, the angles made by the sails and boat axis with the wind could be taken as control variables; having set these, the sailing posture becomes determinate and the boat's velocity is calculable; by choosing the upwind component of this velocity as the performance index to be maximized, a simple optimization problem is formulated. Systems of this type will be called *static systems*.

In later chapters, a more complex situation will be analysed. It will be supposed that the control variables are permitted to vary with the time t in any manner and that this causes the system to move continuously through a succession of states. The quality of the overall response of the system during this period of motion will be taken to be measured by an index C, but C will no longer be a simple function of the u_r, since it will depend upon the infinity of values taken by each control variable during its interval of variation. In these circumstances, C is said to be a *functional* of the functions $u_r(t)$. Systems of this type will be termed *dynamic systems*.

1.2 Unconstrained control

We are supposing that C is a known function of the control variables, so that we can write

$$C = C(u_1, u_2, \ldots, u_N). \quad (1.2.1)$$

This function will be assumed to possess continuous first and second order partial derivatives. The variables u_r may be thought of as the components of a vector in an N-dimensional Euclidean space (the *control space*); this vector will be denoted by u and will be termed the *control vector*. We shall employ the usually accepted notation and exhibit the components of u as the elements of a column matrix, thus:

$$u = \begin{bmatrix} u_1 \\ u_2 \\ \cdot \\ \cdot \\ \cdot \\ u_N \end{bmatrix}. \quad (1.2.2)$$

If the column is not displayed, it will be convenient to write it in the form $(u_1, u_2, \ldots, u_N)^T$, the superscript T implying that the matrix to which it is attached is to be transposed. Equation (1.2.1) will frequently be abbreviated to read $C = C(u)$.

We shall suppose that the values which can be taken by the u_r are all real, but are not otherwise restricted in any way. The control vector u is then said to be *unconstrained*. In future, unless otherwise stated, $u_1, u_2, \ldots, u_N$ will denote the optimal values of the variables, i.e. the values which maximize or minimize C, depending upon the nature of the problem. To derive necessary conditions to be satisfied by these optimal values, it will be necessary to examine the values taken by C at neighbouring points in the control space; such points will be taken to have coordinates $v_r = u_r + \epsilon \xi_r$. If, now, the ξ_r are given arbitrary fixed values and ϵ is regarded as a variable, $C(u + \epsilon \xi)$ will be a function of ϵ having a minimum (or a maximum) at $\epsilon = 0$. It follows from a well-known theorem from the theory of functions of a single variable that

$$\frac{dC}{d\epsilon} = 0, \quad \frac{d^2C}{d\epsilon^2} \geqslant 0 \qquad (1.2.3)$$

at $\epsilon = 0$. (N.B. For simplicity of statement it will be assumed in future, unless otherwise stated, that C is to be minimized; it will be left for the reader to reverse the appropriate inequalities in case C is to be maximized.) Conditions (1.2.3) are clearly valid for all sets of values of the ξ_r.

Assuming that the function C possesses its differentiability and continuity properties over a domain of the control space including the optimal point, it follows that, for sufficiently small ϵ,

$$\frac{dC}{d\epsilon} = \frac{\partial C}{\partial v_r} \frac{dv_r}{d\epsilon} = \frac{\partial C}{\partial v_r} \xi_r; \qquad (1.2.4)$$

the summation with respect to r over its values $1, 2, \ldots, N$ is here indicated by repetition of the index (the repeated index summation convention will be operative throughout the remainder of this text unless otherwise stated). Putting $\epsilon = 0$ in equation (1.2.4), since then $v_r = u_r$ the first of the conditions (1.2.3) takes the form

$$\frac{\partial C}{\partial u_r} \xi_r = 0. \qquad (1.2.5)$$

But this condition must be satisfied for all values of the ξ_r and it accordingly follows that

$$\frac{\partial C}{\partial u_r} = 0. \qquad (1.2.6)$$

A second differentiation of C with respect to ϵ and the setting of ϵ equal to zero yields the result

$$\frac{d^2C}{d\epsilon^2} = \frac{\partial^2 C}{\partial u_r \, \partial u_s} \xi_r \xi_s. \tag{1.2.7}$$

Repetition of both indices r, s implies that a double summation must be carried out on the right-hand member of this equation, which is accordingly a quadratic form in the variables ξ_r. The second of the conditions (1.2.3) requires that this form must be positive semi-definite, i.e. may vanish for a non-null set of values of the ξ_r, but is negative for no set of values of these parameters. A necessary and sufficient condition for this to be true is that the eigenvalues of the symmetric $N \times N$ matrix $A = (a_{rs})$, where $a_{rs} = \partial^2 C / \partial u_r \partial u_s$, should be positive or zero (the eigenvalues of A are defined to be the values of α for which the determinant of $A - \alpha I$ vanishes; I is the unit $N \times N$ matrix).

The point $v = (v_1, v_2, \ldots, v_N)$ will be said to belong to a *neighbourhood* of the optimal point of radius δ if

$$\sqrt{[(v_1 - u_1)^2 + (v_2 - u_2)^2 + \cdots + (v_N - u_N)^2]} < \delta. \tag{1.2.8}$$

Suppose the quadratic form (1.2.7) is positive definite, i.e. positive and non-vanishing for all non-null sets of values of the ξ_r. Then, the eigenvalues of the matrix A will all be positive and non-zero. Since we are assuming the partial derivatives $\partial^2 C / \partial v_r \partial v_s$ to be continuous over a domain containing the optimal point, it will be possible to find a neighbourhood Δ of this point of radius δ within which the eigenvalues associated with the quadratic form $(\partial^2 C / \partial v_r \partial v_s) \xi_r \xi_s$ are also positive and non-vanishing (these eigenvalues being continuous functions of the elements a_{rs} of A). Thus, within Δ, $(\partial^2 C / \partial v_r \partial v_s) \xi_r \xi_s$ is positive definite. Let $u + \eta$ be any point of Δ; then, by Taylor's theorem and using equations (1.2.6),

$$C(u + \eta) = C(u) + \frac{1}{2} \frac{\partial^2 C}{\partial v_r \, \partial v_s} \eta_r \eta_s, \tag{1.2.9}$$

where, in the second derivatives, $v = u + \theta\eta$ ($0 < \theta < 1$). Since the quadratic form in the η_r is positive definite, we can now conclude that

$$C(u + \eta) > C(u), \quad u + \eta \in \Delta, \quad \eta \neq 0. \tag{1.2.10}$$

This is to say that C possesses a local minimum at the point u. Thus, we have proved that sufficient conditions for a local minimum are equations (1.2.6) and the positive definiteness of the form (1.2.7).

PROBLEM 1. Calculate the maxima and minima of the function
$$C(x, y) = x^3 - 12xy + y^3 - 63x - 63y.$$

Solution: At a stationary point

$$\frac{\partial C}{\partial x} = 3x^2 - 12y - 63 = 0, \qquad (1.2.11)$$

$$\frac{\partial C}{\partial y} = -12x + 3y^2 - 63 = 0. \qquad (1.2.12)$$

Hence, $\qquad 3x^2 - 12y = -12x + 3y^2,$

or $\qquad (x - y)(x + y) = 4(y - x).$

Thus, (i) $x = y$ or (ii) $x + y + 4 = 0$. In the first case, substitution for y in equation (1.2.11) leads to $x^2 - 4x - 21 = 0$; i.e. $x = -3, 7$, yielding stationary points $(-3, -3)$, $(7, 7)$. In the second case, substitution for y gives $x^2 + 4x - 5 = 0$; thus, $x = 1, -5$ and the stationary points are $(1, -5)$, $(-5, 1)$.

To determine the nature of the stationary points, we calculate the second derivatives:

$$\frac{\partial^2 C}{\partial x^2} = 6x, \qquad \frac{\partial^2 C}{\partial x\, \partial y} = -12, \qquad \frac{\partial^2 C}{\partial y^2} = 6y.$$

Hence, at the stationary point $(-3, -3)$

$$A = \begin{bmatrix} -18 & -12 \\ -12 & -18 \end{bmatrix}, \qquad A - \alpha I = \begin{bmatrix} -18 - \alpha & -12 \\ -12 & -18 - \alpha \end{bmatrix}.$$

$|A - \alpha I|$ vanishes when $\alpha = -6, -30$. Since both these eigenvalues are negative, $(-3, -3)$ is a local maximum; we find $C_{\max} = 216$.

Similarly, at the point $(7, 7)$, the eigenvalues are found to be 30, 54 and this point is therefore a local minimum ($C_{\min} = -784$).

At both the remaining stationary points $(1, -5)$, $(-5, 1)$, the eigenvalues are the roots of the equation $\alpha^2 + 24\alpha - 324 = 0$. Since the product of these roots is negative, the eigenvalues have opposite signs and the points are neither maxima nor minima. ●

6 ANALYTICAL METHODS OF OPTIMIZATION

PROBLEM 2. The force exerted by the wind on the sail of a yacht is proportional to the square of the wind velocity and to the sine of the angle made by the direction of the wind with the sail; its direction is normal to the sail. Assuming that the drag on the boat is proportional to the square of its velocity, calculate the course which must be steered and the setting of the sail if the component of the yacht's velocity in the upwind direction is to be a maximum.

Solution: In Fig. 1.1, θ is the angle made by the axis of the yacht and ϕ is the angle made by the plane of the sail with the upward wind direction. The force exerted by the wind on the sail will be taken to be $F = cW^2 \sin \phi$ (we neglect the effect the yacht's motion has on the velocity with which the wind strikes the sail). If D is the drag force opposing the yacht's motion and V is its velocity, then $D = kV^2$. Assuming uniform motion, the force components along the boat axis must balance and, hence,

$$kV^2 = cW^2 \sin \phi \sin(\theta - \phi).$$

It is required to choose θ and ϕ so that $V \cos \theta$ is maximized. Since

$$V^2 \cos^2 \theta = \frac{c}{k} W^2 \sin \phi \sin(\theta - \phi)\cos^2 \theta,$$

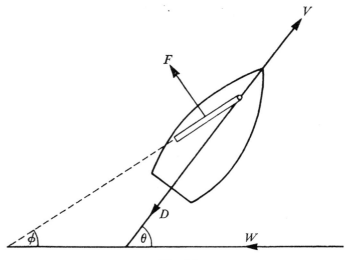

Fig. 1.1

it is more convenient to maximize

$$C = C(\theta, \phi) = \sin \phi \sin(\theta - \phi)\cos^2 \theta.$$

Necessary conditions for a maximum are:

$$\frac{\partial C}{\partial \theta} = \sin \phi [\cos(\theta - \phi)\cos^2 \theta - 2 \sin(\theta - \phi)\cos \theta \sin \theta] = 0,$$

$$\frac{\partial C}{\partial \phi} = [\cos \phi \sin(\theta - \phi) - \sin \phi \cos(\theta - \phi)]\cos^2 \theta = 0.$$

Rejecting the possibility that $\cos \theta$ vanishes, the second of these equations is equivalent to $\tan \phi = \tan(\theta - \phi)$; thus, $\phi = \theta - \phi$, or $\phi = \frac{1}{2}\theta$. Substituting for ϕ in the first equation, it is now easy to show that $\cos \theta = \frac{2}{3}$ (i.e. $\theta = 48°$, $\phi = 24°$ approximately).

For these values of θ and ϕ, it may be verified that the second derivatives of C take the values

$$\frac{\partial^2 C}{\partial \theta^2} = -7, \qquad \frac{\partial^2 C}{\partial \theta \, \partial \phi} = \frac{4}{9}, \qquad \frac{\partial^2 C}{\partial \phi^2} = -\frac{8}{9}.$$

The eigenvalues of the matrix A formed from these elements are both negative. It follows that at least a local maximum of C has been located (it is possible to prove that the maximum is global). ●

1.3 Equality constraints

In this section, we shall suppose that the control vector is subject to one or more equality constraints, i.e. its components u_r are permitted to take only values satisfying equations

$$f_i(u_1, u_2, \ldots, u_N) = c_i, \qquad (1.3.1)$$

where $i = 1, 2, \ldots, M$ and the c_i are constants. If $M = N$, these equations will, normally, possess a finite number of solutions only and if $M > N$, they will be inconsistent; in either case, the optimization problem is empty. We assume, therefore, $M < N$. Any set of values of the u_r which satisfy the constraints (1.3.1) is said to be *admissible* and the corresponding point in control space is termed an *admissible* point.

In this case, in general, it will be possible, in principle, to solve equations (1.3.1) for M of the variables u_r in terms of the remaining $N - M$ (the precise condition for this to be so will be given later).

8   ANALYTICAL METHODS OF OPTIMIZATION

C can then be expressed as a function of $N - M$ variables whose values are unconstrained. By this means, the problem can be reduced to the one already solved. We shall now study the details of this reduction programme.

Firstly, it will be convenient and in conformity with our later nomenclature to denote the M variables for which the constraints are solved by $x_1, x_2, \ldots, x_M$. The remaining independent variables will continue to be represented by $u_1, u_2, \ldots, u_N$. Thus, the constraints (1.3.1) will now be written

$$f_i(x_1, x_2, \ldots, x_M, u_1, u_2, \ldots, u_N) = c_i, \qquad i = 1, 2, \ldots, M, \tag{1.3.2}$$

and the performance index will be given by an equation of the form

$$C = C(x_1, \ldots, x_M, u_1, \ldots, u_N) = C(x, u). \tag{1.3.3}$$

The functions C, f_i will be assumed to possess continuous partial derivatives of the first and second orders with respect to all their arguments over a domain which includes the optimal point within its interior.

As before, u will be referred to as the control vector and $x = [x_1, x_2, \ldots, x_M]^T$ will be termed the *state vector*. The state vector will be regarded as an element in a Euclidean *state space* having M dimensions, the x_i being the coordinates of a point in this space. Since it is being assumed that the x_i are expressible as functions of the u_r, if the point u is allowed to vary in the control space, a consequent variation of the point x occurs in the state space.

Having substituted for the state variables x_i in terms of the control variables u_r in the function $C(x, u)$, let the resulting function of the u_r be denoted by $D(u)$. Then, the values of the control variables which make C a minimum must satisfy the conditions

$$\frac{\partial D}{\partial u_r} = 0, \qquad \frac{\partial^2 D}{\partial u_r \partial u_s} \xi_r \xi_s \geqslant 0, \tag{1.3.4}$$

for arbitrary values of the ξ_r.

Now, the first of these conditions can be expressed in the form

$$\frac{\partial D}{\partial u_r} = \frac{\partial C}{\partial x_i} \frac{\partial x_i}{\partial u_r} + \frac{\partial C}{\partial u_r} = 0, \tag{1.3.5}$$

$r = 1, 2, \ldots, N$. In these equations, the derivatives $\partial x_i / \partial u_r$ are calculable from the constraints; thus, differentiating equations

(1.3.2) partially with respect to u_r (remembering that the x_i can be regarded as functions of the u_r), we obtain

$$\frac{\partial f_i}{\partial x_j}\frac{\partial x_j}{\partial u_r} + \frac{\partial f_i}{\partial u_r} = 0. \qquad (1.3.6)$$

This provides us with M equations for the M unknowns $\partial x_1/\partial u_r$, $\partial x_2/\partial u_r, \ldots, \partial x_M/\partial u_r$ and these have a unique solution provided the determinant $|\partial f_i/\partial x_j|$ of the coefficients does not vanish; this is precisely the condition which must be satisfied if the constraints (1.3.2) are to specify the x_i as functions of the u_r and we shall assume it to be satisfied at the optimal point. Equations (1.3.5), (1.3.6) then provide a set of necessary conditions from which possible optimal sets of values of the u_r can be derived.

These equations can be put in a more convenient and symmetric form by introducing the adjoint equations, namely,

$$\frac{\partial f_i}{\partial x_j}\lambda_i + \frac{\partial C}{\partial x_j} = 0. \qquad (1.3.7)$$

These M equations determine the M quantities λ_i, called the *Lagrange Multipliers* (the non-vanishing of the determinant $|\partial f_i/\partial x_j|$ is again crucial). Multiplying equation (1.3.6) through by λ_i and summing with respect to i, we find

$$\lambda_i \frac{\partial f_i}{\partial x_j}\frac{\partial x_j}{\partial u_r} + \lambda_i \frac{\partial f_i}{\partial u_r} = 0. \qquad (1.3.8)$$

Employing equation (1.3.7), this reduces to the form

$$-\frac{\partial C}{\partial x_j}\frac{\partial x_j}{\partial u_r} + \lambda_i \frac{\partial f_i}{\partial u_r} = 0 \qquad (1.3.9)$$

and, by equation (1.3.5), this in turn is seen to be equivalent to

$$\frac{\partial C}{\partial u_r} + \lambda_i \frac{\partial f_i}{\partial u_r} = 0. \qquad (1.3.10)$$

We next define the *Hamiltonian H* by the equation

$$H = H(x, \lambda, u) = C + \lambda_i f_i. \qquad (1.3.11)$$

Equations (1.3.7), (1.3.10) can then be written in the simple form

$$\frac{\partial H}{\partial x} = 0, \qquad \frac{\partial H}{\partial u_r} = 0. \qquad (1.3.12)$$

This set of equations, together with the constraints (1.3.2), provides us with $(2M + N)$ equations for the same number of unknowns x_i, λ_i, u_r, and, hence, serves to locate all possible-optimal points.

To apply the second necessary condition (1.3.4), we require to calculate the second derivatives $\partial^2 D/\partial u_r \partial u_s$. Differentiating equations (1.3.5), (1.3.6) partially with respect to u_s, the following results are established:

$$\frac{\partial^2 D}{\partial u_r \, \partial u_s} = \frac{\partial C}{\partial x_i} \frac{\partial^2 x_i}{\partial u_r \, \partial u_s} + \frac{\partial^2 C}{\partial x_i \, \partial x_j} \frac{\partial x_i}{\partial u_r} \frac{\partial x_j}{\partial u_s}$$
$$+ \frac{\partial^2 C}{\partial x_i \, \partial u_r} \frac{\partial x_i}{\partial u_s} + \frac{\partial^2 C}{\partial x_i \, \partial u_s} \frac{\partial x_i}{\partial u_r} + \frac{\partial^2 C}{\partial u_r \, \partial u_s}, \quad (1.3.13)$$

$$0 = \frac{\partial f_k}{\partial x_i} \frac{\partial^2 x_i}{\partial u_r \, \partial u_s} + \frac{\partial^2 f_k}{\partial x_i \, \partial x_j} \frac{\partial x_i}{\partial u_r} \frac{\partial x_j}{\partial u_s}$$
$$+ \frac{\partial^2 f_k}{\partial x_i \, \partial u_r} \frac{\partial x_i}{\partial u_s} + \frac{\partial^2 f_k}{\partial x_i \, \partial u_s} \frac{\partial x_i}{\partial u_r} + \frac{\partial^2 f_k}{\partial u_r \, \partial u_s}. \quad (1.3.14)$$

Multiplying equation (1.3.14) through by λ_k, summing over k and adding to equation (1.3.13), it follows that

$$\frac{\partial^2 D}{\partial u_r \, \partial u_s} = \frac{\partial^2 H}{\partial x_i \, \partial x_j} \frac{\partial x_i}{\partial u_r} \frac{\partial x_j}{\partial u_s} + \frac{\partial^2 H}{\partial x_i \, \partial u_r} \frac{\partial x_i}{\partial u_s}$$
$$+ \frac{\partial^2 H}{\partial x_i \, \partial u_s} \frac{\partial x_i}{\partial u_r} + \frac{\partial^2 H}{\partial u_r \, \partial u_s}. \quad (1.3.15)$$

(Note: Use equation (1.3.7).) After calculating the values of the derivatives $\partial x_i/\partial u_r$ from equations (1.3.6), this last equation permits the construction of the quadratic form appearing in the second necessary condition (1.3.4).

PROBLEM 3. A rectangular block is inscribed in the ellipsoid

$$\frac{x^2}{a^2} + \frac{y^2}{b^2} + \frac{z^2}{c^2} = 1. \quad (1.3.16)$$

Determine its maximum volume.

Solution: The eight corners of the block will lie at points $(\pm x, \pm y, \pm z)$, where x, y, z satisfy the equation of the ellipsoid. The volume of the block is given by $V = 8xyz$, where $x \geqslant 0$, $y \geqslant 0$,

$z \geqslant 0$. Hence, xyz is to be maximized subject to the constraint (1.3.16). Any pair of the variables x, y, z can be regarded as control variables and, then, the third is the state variable determined by the constraint.

The Hamiltonian for the problem is

$$H = xyz + \lambda\left(\frac{x^2}{a^2} + \frac{y^2}{b^2} + \frac{z^2}{c^2}\right)$$

and necessary conditions for a maximum are

$$\frac{\partial H}{\partial x} = yz + 2\lambda x/a^2 = 0,$$

$$\frac{\partial H}{\partial y} = zx + 2\lambda y/b^2 = 0,$$

$$\frac{\partial H}{\partial z} = xy + 2\lambda z/c^2 = 0.$$

The first equation yields $x^2/a^2 = -xyz/2\lambda$; y^2/b^2, z^2/c^2 can be found similarly from the other equations. Substitution in the constraint shows that $-3xyz/2\lambda = 1$; i.e. $\lambda = -3xyz/2$. It now follows that $x = a/\sqrt{3}$, $y = b/\sqrt{3}$, $z = c/\sqrt{3}$ and, hence, $\lambda = -abc/2\sqrt{3}$.

Treating z as the state variable and regarding it as a function of x, y, differentiation of the constraints leads to the equations

$$\frac{2x}{a^2} + \frac{2z}{c^2}\frac{\partial z}{\partial x} = \frac{2y}{b^2} + \frac{2z}{c^2}\frac{\partial z}{\partial y} = 0.$$

Thus, at the optimal point, $\partial z/\partial x = -c/a$ and $\partial z/\partial y = -c/b$.

Equation (1.3.15) applied to this problem gives

$$\frac{\partial^2 D}{\partial x \, \partial y} = \frac{\partial^2 H}{\partial z^2}\frac{\partial z}{\partial x}\frac{\partial z}{\partial y} + \frac{\partial^2 H}{\partial x \, \partial z}\frac{\partial z}{\partial y} + \frac{\partial^2 H}{\partial y \, \partial z}\frac{\partial z}{\partial x} + \frac{\partial^2 H}{\partial x \, \partial y} = -\frac{2c}{\sqrt{3}}.$$

Similarly, it is found that $\partial^2 D/\partial x^2 = -4bc/\sqrt{3}a$, $\partial^2 D/\partial y^2 = -4ca/\sqrt{3}b$.

Forming the matrix A, its eigenvalues are found to satisfy the equation

$$\alpha^2 + \frac{4c}{\sqrt{3}}\left(\frac{a}{b} + \frac{b}{a}\right)\alpha + 4c^2 = 0.$$

This, clearly, has negative roots, indicating that V has been maximized. Then, $V_{\max} = 8abc/3\sqrt{3}$. ●

PROBLEM 4. The cost function is given by

$$C = \tfrac{1}{2}p_{ij}x_ix_j + \tfrac{1}{2}q_{rs}u_ru_s.$$

The constraints are linear and given by

$$x_i = b_{ir}u_r + c_i.$$

It is required to minimize C.

Solution: The Hamiltonian is

$$H = \tfrac{1}{2}p_{ij}x_ix_j + \tfrac{1}{2}q_{rs}u_ru_s + \lambda_i(x_i - b_{ir}u_r)$$

and (assuming $p_{ij} = p_{ji}$, $q_{rs} = q_{sr}$) necessary conditions for a minimum are

$$\frac{\partial H}{\partial x_i} = p_{ij}x_j + \lambda_i = 0,$$

$$\frac{\partial H}{\partial u_r} = q_{rs}u_s - \lambda_i b_{ir} = 0.$$

It is convenient to write these as matrix equations, namely,

$$Px + \lambda = 0, \qquad Qu - B^T\lambda = 0,$$

where x, λ, u are columns, P is the matrix whose ijth element is p_{ij}, Q has rsth element q_{rs} and B has irth element b_{ir}. P, Q are symmetric square matrices of orders M, N respectively and B is of type $M \times N$.

Writing the constraints in the matrix form $x = Bu + c$, we can now eliminate x and λ to yield the equation

$$(Q + B^TPB)u + B^TPc = 0.$$

We shall now assume that $Q + B^TPB$ is positive definite and, hence, that the inverse $(Q + B^TPB)^{-1}$ exists. Thus,

$$u = -(Q + B^TPB)^{-1}B^TPc.$$

x and λ can now be found in the forms

$$x = [I - B(Q + B^TPB)^{-1}B^TP]c,$$
$$\lambda = [PB(Q + B^TPB)^{-1}B^TP - Pc].$$

STATIC SYSTEMS

It follows from the constraints that $\partial x_i/\partial u_r = b_{ir}$. Equation (1.3.15) applied to this problem accordingly gives the result

$$\frac{\partial^2 D}{\partial u_r \partial u_s} = p_{ij}b_{ir}b_{js} + q_{rs}.$$

The condition for a minimum is that the quadratic form

$$\frac{\partial^2 D}{\partial u_r \partial u_s}\xi_r\xi_s = p_{ij}b_{ir}b_{js}\xi_r\xi_s + q_{rs}\xi_r\xi_s = \xi^T(B^T P B + Q)\xi$$

must be positive definite, and this we have already assumed to be the case.

Finally,

$$\begin{aligned}C_{\min} &= \tfrac{1}{2}x^T P x + \tfrac{1}{2}u^T Q u \\ &= -\tfrac{1}{2}x^T\lambda + \tfrac{1}{2}u^T B^T \lambda \\ &= -\tfrac{1}{2}(x - Bu)^T\lambda \\ &= -\tfrac{1}{2}c^T\lambda \\ &= \tfrac{1}{2}c^T[P - PB(Q + B^T P B)^{-1}B^T P]c.\end{aligned}$$ ●

1.4 The Lagrange multipliers

Suppose that the constants c_i appearing in the constraints (1.3.2) have their values altered slightly to $c_i + \delta c_i$. A new optimization problem then arises. Let the optimal values of the state and control variables for this new problem be denoted by $x_i + \delta x_i$, $u_r + \delta u_r$. These must satisfy the equations

$$f_i(x + \delta x, u + \delta u) = c_i + \delta c_i \qquad (1.4.1)$$

and, working to the first order of small quantities, this equation is equivalent to

$$\frac{\partial f_i}{\partial x_j}\delta x_j + \frac{\partial f_i}{\partial u_r}\delta u_r = \delta c_i; \qquad (1.4.2)$$

(remember $f_i(x, u) = c_i$). Multiplying this last equation through by λ_i and summing over i, we get

$$\lambda_i\frac{\partial f_i}{\partial x_j}\delta x_j + \lambda_i\frac{\partial f_i}{\partial u_r}\delta u_r = \lambda_i\delta c_i. \qquad (1.4.3)$$

Using equations (1.3.7), (1.3.10), this can be written

$$-\frac{\partial C}{\partial x_j}\delta x_j - \frac{\partial C}{\partial u_r}\delta u_r = \lambda_i \delta c_i. \quad (1.4.4)$$

To emphasize that optimal values are being taken everywhere, a subscript zero will be attached to C. Then, if C_0 denotes the optimal cost for the original problem and $C_0 + \delta C_0$ the optimal cost for the new problem, equation (1.4.4) indicates that

$$\delta C_0 = -\lambda_i \delta c_i. \quad (1.4.5)$$

Thus, in the limit as $\delta c_i \to 0$, we have

$$\frac{\partial C_0}{\partial c_i} = -\lambda_i. \quad (1.4.6)$$

This equation reveals the mathematical significance of the Lagrange multipliers.

1.5 Inequality constraints

Suppose that $C(u)$ is to be minimized over all vectors u satisfying conditions

$$f_i(u) \geqslant c_i, \quad i = 1, 2, \ldots, M. \quad (1.5.1)$$

The functions $f_i(u)$ will be assumed to possess continuous second order partial derivatives throughout the control space. Admissible points in this space will determine a closed region R (i.e. including the boundary points) bounded by the hypersurfaces $f_i = c_i$.

If the optimal point is an interior point of R, none of the constraints is operative in a sufficiently small neighbourhood of this point and they can accordingly be ignored. The conditions derived in section 1.2 are then still applicable.

If, however, the optimal point O is a boundary point lying on the first $Q(\leqslant N)$ hypersurfaces $f_m = c_m$ ($m = 1, 2, \ldots, Q$), but not on the remainder, the constraints $f_i \geqslant c_i$ ($i = 1, 2, \ldots, Q$) will now be operative but the constraints $f_i \geqslant c_i$ ($i = Q + 1, \ldots, M$) will still be ineffective over a sufficiently small neighbourhood Δ of O. We shall therefore study the *associated problem* of minimizing C subject to the equality constraints $f_m = c_m$ ($m = 1, 2, \ldots, Q$). We shall refer to the original problem as $\mathscr{P}$ and the associated problem as $\mathscr{P}'$. Then, the optimal point O for $\mathscr{P}$ must be, at least, a local minimum for $\mathscr{P}'$ and it follows that the necessary conditions derived in section 1.3 for the class of problems of the type $\mathscr{P}'$ must be satisfied at O.

Let C_0 be the value of C at O and let λ_m be the multipliers arising in the solution of $\mathscr{P}'$. Consider a third problem $\mathscr{P}''$ in which C is to be minimized subject to the perturbed constraints $f_m = c_m + \delta c_m$ ($\delta c_m \geqslant 0$). Assuming the coordinates of O are continuously dependent on the c_m, if the δc_m are sufficiently small, a point O' lying in Δ can be found which is a local minimum for $\mathscr{P}''$. Since $O' \in \Delta$, the constraints $f_i \geqslant c_i$ ($i = Q+1, \ldots, M$) are all satisfied at O'. Further, since the δc_m are positive, the remaining constraints $f_m \geqslant c_m$ are also satisfied at O'. Hence, $O' \in R$. Let C_0, $C_0 + \delta C_0$ be the values of the cost at O, O' respectively. Then, since O is the optimal point in R, $\delta C_0 \geqslant 0$. But, by equation (1.4.5), $\delta C_0 = -\lambda_m \delta c_m$, to the first order in the δc_m. Since the δc_m are arbitrary positive quantities, we now conclude that

$$\lambda_m \leqslant 0, \quad m = 1, 2, \ldots, Q. \tag{1.5.2}$$

It has now been proved that, at an optimal point O where the first Q of the constraints (1.5 1) are effective (i.e. O lies on the hypersurfaces $f_m = c_m$, $m = 1, 2, \ldots, Q$), the following conditions are necessarily satisfied.

$$\frac{\partial C}{\partial u_r} + \lambda_m \frac{\partial f_m}{\partial u_r} = 0, \quad \lambda_m \leqslant 0, \quad f_m = c_m. \tag{1.5.3}$$

Putting

$$H = C + \lambda_i f_i, \tag{1.5.4}$$

these conditions can be written in the alternative form

$$\left.\begin{aligned}\frac{\partial H}{\partial u_r} &= 0, \\ \lambda_i &\leqslant 0 \text{ if } f_i(u) = 0, \\ \lambda_i &= 0 \text{ if } f_i(u) > 0.\end{aligned}\right\} \tag{1.5.5}$$

We have just established necessary conditions to be satisfied at O. By the usual slight strengthening of these conditions, they become sufficient conditions for O to be a local minimum. Thus, suppose it is known that O is a local minimum for the problem $\mathscr{P}'$ and that $\lambda_m < 0$ ($m = 1, 2, \ldots, Q$); sufficient conditions for the validity of the first part of this hypothesis have been given in section 1.3; the second part is the strengthened condition. Let P be any point belonging to $\Delta \cap R$ and denote the values of $f_m(u)$ at P by $c_m + \delta c_m$; then $\delta c_m \geqslant 0$. Let $C_0 + \delta C$ be the cost at P. Then P is an admissible point for the problem $\mathscr{P}''$ and, since O' is a local minimum for this problem, it follows that $C_0 + \delta C \geqslant C_0 + \delta C_0$, i.e. $\delta C \geqslant \delta C_0$. But, to the first order, $\delta C_0 = -\lambda_m \delta c_m$ and, therefore, $\delta C \geqslant -\lambda_m \delta c_m$. If all the λ_m are negative and non-zero, we can now conclude that $\delta C > 0$, except possibly when the δc_m all vanish. In this latter event, however, P is an admissible point for $\mathscr{P}'$ and our assumption that O is a local minimum for this problem then gives $\delta C \geqslant 0$. Thus, in all circumstances, $\delta C \geqslant 0$ and this proves that O is a local minimum.

If the original problem is amended by reversal of the inequalities appearing in some or all of the constraints (1.5.1), the corresponding inequalities satisfied by the λ_m must also be reversed. If the problem is amended by the omission of equalities from all of these constraints, all boundary points become *ipso facto* inadmissible. Hence, if an optimal point exists, it must be an interior point at which the conditions derived in section 1.2 apply. However, in such a case, it is possible that the original problem $\mathscr{P}$ has a boundary point solution and, hence, that C can be made to approach this minimum value as close as we please without violating the amended constraints. In these circumstances, the amended problem has no solution.

PROBLEM 5. Maximize xyz subject to the constraint

$$\frac{x^2}{a^2} + \frac{y^2}{b^2} + \frac{z^2}{c^2} \leqslant 1.$$

Solution: Clearly, no interior point of the ellipsoidal region R can be a maximum point for xyz. We therefore assume the point required lies on the boundary of R and solve the associated problem. This has already been done in Problem 3, where it was shown that a local maximum occurs on the ellipsoid at the point $(a/\sqrt{3}, b/\sqrt{3}, c/\sqrt{3})$; the associated value of the multiplier λ was calculated to be $-abc/2\sqrt{3}$. This is negative, indicating that this point is also a local maximum for the present problem (Note: C is being maximized and the inequality constraint is the reverse of that assumed in equation (1.5.1).) ●

1.6 Linear programming

A special case of the type of problem studied in the last section is of great practical importance. This is when

$$C = a_r u_r \tag{1.6.1}$$

is to be minimized subject to the constraints

$$b_{ir} u_r \geqslant c_i, \quad i = 1, 2, \ldots, M. \tag{1.6.2}$$

In this case the boundary hypersurfaces of the region of admissible points in the control space are hyperplanes.

If an interior point of R were minimal, $\partial C/\partial u_r \, (= a_r)$ would have to vanish for all values of r, i.e. C would vanish identically. Rejecting this case, the minimal point, if it exists, must lie on the boundary. Suppose it lies on all the hyperplanes

$$b_{mr} u_r = c_m, \quad m = 1, 2, \ldots, Q. \tag{1.6.3}$$

Since, in general, more than N such hyperplanes will have no point in common, we shall assume $Q \leqslant N$. The Hamiltonian for the associated problem is given by

$$H = a_r u_r + \lambda_m b_{mr} u_r \tag{1.6.4}$$

and, at the minimal point, it is necessary that

$$\frac{\partial H}{\partial u_r} = a_r + b_{mr} \lambda_m = 0, \tag{1.6.5}$$

$r = 1, 2, \ldots, N$. If $Q < N$, there are insufficient unknowns λ_m to permit these equations to be satisfied (we consider only the general case). Hence $Q = N$ and the minimal point must lie at the intersection of exactly N of the boundary hyperplanes. Since $Q \leqslant M$, this is only possible if $M \geqslant N$; if $M < N$, the problem has no solution.

Assuming, therefore, $Q = N \leqslant M$, the sets of equations (1.6.3), (1.6.5) can be solved for u, λ respectively. For the point u to be minimal, the λ_r must all be negative. If they are not, another set of N boundary hyperplanes must be selected to replace the set (1.6.3) and u, λ recalculated. If a minimal point exists, a finite number of trials of this type will locate it.

If the constraints (1.6.2) are augmented by the requirement that the variables u_r should not be negative, i.e. $u_r \geqslant 0, r = 1, 2, \ldots, N$, we have what is termed a *linear programming problem*.

If this problem has a solution, the minimal point will be at the intersection of N hyperplanes, which can now include any of the coordinate hyperplanes $u_r = 0$. Suppose that the nomenclature is chosen such that the minimal point lies on the hyperplanes $b_{ir}u_r = c_i$ ($i = 1, 2, \ldots, Q \leqslant M$) and the coordinate hyperplanes $u_{Q+1} = u_{Q+2} = \cdots = u_N = 0$. Thus, at the minimal point,

$$b_{mn}u_n = c_m, \quad b_{pn}u_n > c_p, \quad u_l = 0, \quad u_n > 0, \quad (1.6.6)$$

where $m, n = 1, 2, \ldots, Q$, $p = Q + 1, Q + 2, \ldots, M$ and $l = Q + 1, Q + 2, \ldots, N$.

The Hamiltonian for the associated problem is given by

$$H = a_r u_r + \lambda_m b_{mr} u_r + \Lambda_l u_l, \quad (1.6.7)$$

where, for minimality, we must have

$$\lambda_m < 0, \quad \Lambda_l < 0. \quad (1.6.8)$$

The necessary conditions $\partial H/\partial u_r = 0$ separate into two groups, (i) $r = 1, 2, \ldots, Q$ and (ii) $r = Q + 1, Q + 2, \ldots, N$, namely

$$a_n + \lambda_m b_{mn} = 0, \quad (1.6.9)$$

$$a_l + \lambda_m b_{ml} + \Lambda_l = 0. \quad (1.6.10)$$

Solution of the equations (1.6.9) will yield the multipliers λ_m and equations (1.6.10) then give the multipliers Λ_l. These multipliers must all be negative for a minimal point.

The *dual linear programming problem* is that of maximizing

$$C' = c_i u_i' \quad (1.6.11)$$

subject to the constraints

$$b_{ir} u_i' \leqslant a_r, \quad (1.6.12)$$

$$u_i' \geqslant 0. \quad (1.6.13)$$

It will now be proved that, provided equations (1.6.9), (1.6.10) correspond to a minimal point for the original linear programming problem a maximal point P for the dual problem is determined by the equations

$$u_m' = -\lambda_m, \; u_{Q+1}' = u_{Q+2}' = \cdots = u_M' = 0. \quad (1.6.14)$$

STATIC SYSTEMS

Firstly, using equations (1.6.9), (1.6.10) we note that

$$b_{ir}u_i' = -b_{mr}\lambda_m = a_r + \Lambda_q, \qquad (1.6.15)$$

where we put $\Lambda_r = 0$ for $r = 1, 2, \ldots, Q$. Then, since Λ_r vanishes for $r = 1, 2, \ldots, Q$ and Λ_r is negative for $r = Q + 1, Q + 2, \ldots, N$, the first set of constraints (1.6.12) is clearly satisfied. Also, since the λ_m are negative, the second set (1.6.13) is also satisfied. Thus, the point P defined by equations (1.6.14) is admissible.

Putting $r = n = 1, 2, \ldots, Q$ in equation (1.6.15), we see that P lies on the hyperplanes $b_{in}u_i' = a_n$; it also lies on the coordinate hyperplanes $u_p' = 0, p = Q + 1, Q + 2, \ldots, M$. Thus, the associated problem has Hamiltonian

$$H' = c_i u_i' + \lambda_n' b_{in} u_i' + \Lambda_p' u_p', \qquad (1.6.16)$$

with multipliers λ_n', Λ_p'. Necessary conditions to be satisfied at a maximal point are $\partial H'/\partial u_i' = 0$, i.e.

$$c_m + \lambda_n' b_{mn} = 0, \qquad (1.6.17)$$

$$c_p + \lambda_n' b_{pn} + \Lambda_p' = 0. \qquad (1.6.18)$$

Comparing equations (1.6.17) with the first set of equations (1.6.6), we deduce that $\lambda_n' = -u_n$. It follows that $\lambda_n' < 0$. Also, equations (1.6.18) give

$$\Lambda_p' = -c_p - \lambda_n' b_{pn} = -c_p + b_{pn} u_n > 0, \qquad (1.6.19)$$

using the inequalities (1.6.6). Hence the multipliers λ_n', Λ_p' satisfy the sufficiency conditions for a maximal point and the required result is proved.

Further, if we write equations (1.6.6), (1.6.9) in the matrix form

$$Bu = c, \qquad B^T \lambda = -a, \qquad (1.6.20)$$

where B is the $Q \times Q$ matrix with elements b_{mn} and u, a, c are the columns $[u_1, u_2, \ldots, u_Q]^T$, $[a_1, a_2, \ldots, a_Q]^T$, $[c_1, c_2, \ldots, c_Q]^T$ respectively, then

$$u = B^{-1}c, \qquad \lambda = -(B^T)^{-1}a. \qquad (1.6.21)$$

Hence

$$u' = -\lambda = (B^T)^{-1}a. \qquad (1.6.22)$$

It follows that

$$C_{\min} = a_m u_m = a^T u = a^T B^{-1} c. \qquad (1.6.23)$$

$$C'_{\max} = c_m u_m' = u'^T c + a^T B^{-1} c. \qquad (1.6.24)$$

Thus $C_{\min} = C'_{\max}$.

It may be proved quite generally that a linear programming problem has a solution if, and only if, the dual problem has a solution and that, if the solutions exist, the corresponding minimum and maximum values are identical.

PROBLEM 6. A manufacturer produces metal boxes and keys. To produce a box, a lathe must operate for 1 minute, a grinder for 3 minutes and a drill for 3 minutes. To produce a key, the respective times are 2 minutes, 1 minute and $1\tfrac{1}{2}$ minutes. The available weekly machine capacities in minutes are as follows: lathe 40,000, grinder 45,000, drill 48,000. He makes a profit of 10p per box and 15p per key. How many boxes and keys should be produced during a week to maximize his profit.

Solution: Let x, y thousand be the optimal weekly number of boxes and keys respectively. Then the weekly operating time for a lathe is $x + 2y$ (in units of one thousand minutes). Since the total lathe capacity is 40 units this leads to the inequality

$$x + 2y \leqslant 40.$$

Similarly, we derive the inequalities

$$3x + y \leqslant 45, \qquad 3x + \tfrac{3}{2}y \leqslant 48.$$

His net profit is $1000C$, where

$$C = 10x + 15y$$

and this is to be maximized subject to the above inequality constraints and, also $x \geqslant 0, y \geqslant 0$.

In the xy-plane, the boundary lines for the inequality constraints are the straight lines AB, CD, EF (Fig. 1.2). Thus, the point (x, y) is constrained to lie inside or on the polygon $OCQPB$. We know from theory that the maximal point must be one of the vertices of this figure. By calculation, we find that the coordinates of these vertices are as follows: $C(15, 0)$, $Q(13, 6)$, $P(8, 16)$, $B(0, 20)$. At these points, the values of C are 150, 220, 320, 300 respectively.

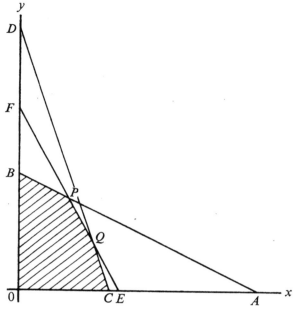

Fig. 1.2

Thus, the maximum profit is 320,000p and occurs when 8,000 boxes and 16,000 keys are made. ●

PROBLEM 7. Formulate the dual of the previous problem and solve it.

Solution: The dual problem is to minimize

$$C' = 40u + 45v + 48w$$

subject to the constraints

$$u + 3v + 3w \geqslant 10,$$
$$2u + v + \tfrac{3}{2}w \geqslant 15,$$
$$u \geqslant 0, \quad v \geqslant 0, \quad w \geqslant 0.$$

From the 5 boundary planes in 3-dimensional uvw-space, 10 triads can be selected; each triad of planes has a point in common, the 10 such points having coordinates (20/3, 0, 10/9), (7, 1, 0), (10, 0, 0),

(0, 0, 10), (0, 15, 0), (0, −20, 70/3), (0, 0, 10/3), (0, 10/3, 0), (15/2, 0, 0), (0, 0, 0). Only the first five of these points are admissible; the remainder fail to satisfy the constraints. Calculating C' at the admissible points, we obtain the values 320, 325, 400, 480, 675. Thus $C'_{min} = 320$, the optimal variable values being $u = 20/3$, $v = 0$, $w = 10/9$.
As expected $C'_{min} = C_{max}$. ●

Exercises 1

1. Show that the function
$$6x^2 + 8y^3 + 3(2x + y - 1)^2$$
possesses a local minimum value of $11/16$.

2. Identical particles are placed at the n points (x_i, y_i) $(i = 1, 2, \ldots, n)$ in the plane of rectangular axes Oxy. The straight line
$$x \cos \theta + y \sin \theta = p$$
is to be constructed so that the moment of inertia of the system of particles about it is a minimum. Show that p and θ must be chosen to satisfy the equations
$$p = \bar{x} \cos \theta + \bar{y} \sin \theta, \qquad \tan 2\theta = \frac{2H}{A - B},$$
where
$$\bar{x} = \frac{1}{n} \sum x_i, \qquad \bar{y} = \frac{1}{n} \sum y_i,$$
$$A = \frac{1}{n} \sum x_i^2 - \bar{x}^2, \qquad B = \frac{1}{n} \sum y_i^2 - \bar{y}^2,$$
$$H = \frac{1}{n} \sum x_i y_i - \bar{x}\bar{y}.$$

3. The geometric mean of n positive quantities is fixed. Show that their arithmetric mean is a minimum when they are equal.

4. If x, y, z are subject to the constraints
$$x + y + z = 1, \qquad x^2 + y^2 + z^2 = 1,$$
show that $x^3 + y^3 + z^3$ has a minimum value $8/27$ and a maximum value 1.

5. Relative to rectangular axes $Oxyz$, A is the point (a, b, c). Write down equations from which can be calculated the coordinates of a point P lying on the surface $f(x, y, z) = 0$, if it is given that the distance AP is stationary with respect to variation of P. Deduce that the line AP has direction ratios $\partial f/\partial x$, $\partial f/\partial y$, $\partial f/\partial z$. Show that, if the surface is the sphere $x^2 + y^2 + z^2 = r^2$, P has two possible positions with coordinates $x = \pm ar/(a^2 + b^2 + c^2)^{\frac{1}{2}}$, etc.

6. Referring to problem 4, suppose the cost function is taken to be

$$C = \tfrac{1}{2} p_{ij} x_i x_j + \tfrac{1}{2} q_{rs} u_r u_s + r_{ir} x_i u_r$$

and it is assumed that the matrix $B^T P B + Q + B^T R + R^T B$ is positive definite. Show that C can be minimized by taking

$$u = (B^T P B + Q + B^T R + R^T B)^{-1}(B^T P + R^T)c$$

and that

$C_{\min}$
$= \tfrac{1}{2} c^T [(PB + R)(B^T P B + Q + B^T R + R^T B)^{-1}(B^T P + R^T) + P]c.$

7. If the equality constraints in Ex. 4 above are replaced by the constraints

$$x + y + z \geqslant 1, \qquad x^2 + y^2 + z^2 \geqslant 1,$$

show that the points at which $x^3 + y^3 + z^3$ takes the value 8/27 remain minima. Show also that if the equality constraints are replaced by the constraint $x^2 + y^2 + z^2 \leqslant 1$, the points at which $x^3 + y^3 + z^3$ takes the value $+1$ remain maxima.

8. x, y, z are positive or zero quantities. Maximize $4x - 2y - z$ subject to the constraints $x + y + z \leqslant 3$, $2x + 2y + z \leqslant 4$, $x - y \leqslant 0$. (Ans. Maximum value 2 occurs at $x = y = 1$, $z = 0$.)

2 | CONTROL SYSTEMS

2.1 Definitions

We shall now transfer attention from static to dynamic systems, i.e. to systems whose configuration varies with the time t as it responds to controlled changes in certain quantities u_r ($r = 1, 2, \ldots, N$). These quantities will be termed the *control variables* or the *input variables* for the system and will be regarded as the components of a vector u in an N-dimensional control space. If U is the point in control space having coordinates u_r with respect to a rectangular frame, the locus of U as t varies will be called the *control trajectory*.

It will be assumed that the configuration or state of the system at any instant can be specified by giving the values of M quantities x_i ($i = 1, 2, \ldots, M$) called the *state variables*. The locus of the point X having coordinates x_i in the state space will be referred to as the *state trajectory*.

It is a defining property of the state variables that if their values are known at any instant $t = \tau$ and if $u(t)$ is given for $t \geqslant \tau$, then $x(t)$ must be calculable for $t \geqslant \tau$, i.e. knowledge of the system's state at any instant τ and of the subsequent manner of control, implies that we shall be able to predict the system's subsequent behaviour. More particularly, it will be assumed that the system's behaviour is governed by a set of M first order differential equations

$$\dot{x}_i = f_i(x_i, \ldots, x_M, u_1, \ldots, u_N, t), \tag{2.1.1}$$

which will be called the *state equations*. This set of equations will frequently be abbreviated to the form

$$\dot{x} = f(x, u, t), \tag{2.1.2}$$

x, f, u referring to the associated column matrices.

It will be assumed that the control variables $u_r(t)$ are continuous functions of t for all values of t in the interval $t^0 \leqslant t \leqslant t^1$ (except that, later, we shall permit these functions to have a finite number of finite discontinuities, i.e. $u_r(t)$ will then be *piecewise continuous*).

$t = t^0$ will be called the *initial instant* and $t = t^1$ the *final instant*. It will also be assumed that the functions f_i possess continuous second order partial derivatives with respect to all variables x_i, u_r, t over a sufficiently extensive region to validate subsequent arguments. It will then follow from the fundamental existence theorem for the system of first order differential equations (2.1.1) that, if the initial state $x_i = x_i^0$ of the system at $t = t^0$ is given, the set of equations (2.1.1) determines the functions $x_i(t)$ uniquely in the interval $[t^0, t^1]$ and that these functions will have continuous derivatives of the first three orders, except at instants when the control variables are discontinuous; at such instants, the state variables will be continuous, but their derivatives will, in general, be discontinuous.

It may happen that the equations governing the behaviour of a control system are not expressed originally in the form (2.1.1). In particular, second and higher order derivatives of some of the state variables may occur. In these circumstances, the x_i will not constitute a complete set of state variables; since a knowledge of their values at some time $t = \tau$ will be insufficient to permit us to calculate the subsequent response of the system to a given control $u(t)$; the values at $t = \tau$ of some derivatives of the x_i will also be required. It will then be necessary to introduce additional state variables until an equivalent system of equations has been derived having the structure exhibited at (2.1.1). For example, if the control system is governed by a single second order equation

$$\ddot{x} = f(x, \dot{x}, u, t), \tag{2.1.3}$$

by treating $\dot{x}$ as a further state variable y, we obtain the equivalent system

$$\left. \begin{array}{l} \dot{x} = y \\ \dot{y} = f(x, y, u, t). \end{array} \right\} \tag{2.1.4}$$

It may sometimes happen that derivatives of the controlling quantities are present in the equations governing the system's behaviour. Such derivatives must be eliminated by transformation of variables before the state and control variables can be correctly identified and the techniques described in this book applied (e.g. see Problem 10).

When the equations governing the behaviour of a control system have been reduced to the form (2.1.1), they will be said to be in *canonical form*.

Although the response of a control system is described in terms of its state variables, these quantities are chosen for their convenience for this purpose and will very often not be directly observable. Our object in controlling the system may be to cause yet other quantities to vary in a manner we regard as desirable. Such quantities will be termed *output variables* and they will be supposed to be directly observable, e.g. rotations of shafts, temperatures of radiators, etc. which it is our purpose to control. The output quantities will be denoted by p_l, $l = 1, 2, \ldots, P$ and it will be assumed that their values are calculable at a time t when the state and control vectors are known, thus:

$$p_l = p_l(x, u, t). \tag{2.1.5}$$

2.2 Linear control systems

If the functions f_i appearing in the state equations (2.1.1) are linear and homogeneous in the state and control variables, the control system is said to be *linear*. These equations can then be expressed in the form

$$\dot{x}_i = a_{ij}x_j + b_{ir}u_r, \tag{2.2.1}$$

where the coefficients a_{ij}, b_{ir} are, in general, functions of t. If the functions f_i are not explicitly dependent upon t, the system is said to be *autonomous*. Thus, if the linear system is autonomous, the coefficients a_{ij}, b_{ir} will be constants.

To solve the equations (2.2.1) in the general case when the coefficients are functions of t, it is helpful firstly to study the special case when the system runs free from control, i.e. when $u = 0$. Then,

$$\dot{x}_i = a_{ij}x_j, \tag{2.2.2}$$

or, in matrix form,

$$\dot{x} = Ax. \tag{2.2.3}$$

We shall denote by $x_i = \phi_{ik}(t, t^0)$ the unique solution of the system of equations (2.2.2) which satisfies the initial conditions $x_i(t^0) = \delta_{ik}$, where δ_{ik} is the Kronecker delta symbol. Then, the solution of equations (2.2.2) satisfying the initial conditions $x_i(t^0) = x_i^0$ is given by

$$x_i(t) = x_k^0 \phi_{ik}(t, t^0). \tag{2.2.4}$$

This follows since, then,

$$\dot{x}_i = x_k^0 \frac{\partial \phi_{ik}}{\partial t} = x_k^0 a_{ij} \phi_{jk} = a_{ij} x_j, \qquad (2.2.5)$$

and $x_i(t^0) = x_k^0 \delta_{ik} = x_i^0$.

The $M \times M$ matrix whose ikth element is ϕ_{ik} will be denoted by $\phi(t, t^0)$. Since

$$\frac{\partial \phi_{ik}}{\partial t} = a_{ij} \phi_{jk},$$

we have equivalently

$$\frac{\partial \phi}{\partial t} = A\phi; \qquad (2.2.6)$$

i.e. ϕ satisfies the equation (2.2.3). Also;

$$\phi(t^0, t^0) = I, \qquad (2.2.7)$$

where I is the $M \times M$ unit matrix. Equation (2.2.4) now has the equivalent matrix form

$$x(t) = \phi(t, t^0) x^0. \qquad (2.2.8)$$

Thus, multiplication of the state vector at $t = t^0$ by the matrix $\phi(t, t^0)$ transforms it into the state vector at time t. ϕ is accordingly termed the *state transition matrix*.

Let t_1, t_2, t_3 be any three values of t. Then, since t, t^0 can be chosen to have any values, it follows from equation (2.2.8) that, for the free-running system,

$$\left.\begin{array}{l} x(t_2) = \phi(t_2, t_1) x(t_1), \\ x(t_3) = \phi(t_3, t_2) x(t_2). \end{array}\right\} \qquad (2.2.9)$$

Eliminating $x(t_2)$, we find that

$$x(t_3) = \phi(t_3, t_2) \phi(t_2, t_1) x(t_1). \qquad (2.2.10)$$

But, we must also have

$$x(t_3) = \phi(t_3, t_1) x(t_1) \qquad (2.2.11)$$

and, hence

$$\phi(t_3, t_1) = \phi(t_3, t_2) \phi(t_2, t_1). \qquad (2.2.12)$$

This is the *multiplicative property* of the state transition matrix.

Putting $t_3 = t_1$ in the identity (2.2.12), it reduces to the form

$$I = \phi(t_1, t_2)\phi(t_2, t_1), \qquad (2.2.13)$$

i.e.
$$\phi(t_2, t_1) = \phi^{-1}(t_1, t_2). \qquad (2.2.14)$$

We can now use the state transition matrix to solve the system of equations (2.2.1) for any given control $u(t)$ and initial state $x(t^0) = x^0$. Writing these equations in the matrix form

$$\dot{x} = Ax + Bu, \qquad (2.2.15)$$

we put
$$x(t) = \phi(t, t^0)y(t), \qquad (2.2.16)$$

where y is a new state vector. Substituting in equation (2.2.15), it is found that

$$\frac{\partial \phi}{\partial t} y + \phi \dot{y} = A\phi y + Bu. \qquad (2.2.17)$$

Equation (2.2.6) now shows that this reduces to

$$\phi \dot{y} = Bu. \qquad (2.2.18)$$

Thus, $\dot{y} = \phi^{-1}Bu$ and, integrating over the interval (t^0, t), we find that

$$y(t) = y^0 + \int_{t^0}^{t} \phi^{-1}(\tau, t)B(\tau)u(\tau)\,d\tau. \qquad (2.2.19)$$

But, $x^0 = \phi(t^0, t^0)y^0 = Iy^0 = y^0$ and, by equation (2.2.14), $\phi^{-1}(\tau, t) = \phi(t, \tau)$. Hence, it follows from equations (2.2.16), (2.2.19) that

$$x(t) = \phi(t, t^0)x^0 + \int_{t^0}^{t} \phi(t, \tau)B(\tau)u(\tau)\,d\tau. \qquad (2.2.20)$$

This is the solution we are seeking.

PROBLEM 8. The response of a system to an input control quantity $u(t)$ is governed by the equation

$$t^2\ddot{x} + 4t\dot{x} + 2x = u.$$

Calculate its state transition matrix. Hence determine its response from an initial state $x = \dot{x} = 0$ at $t = 1$ to the control $u = 1$.

Solution. Introducing a further state variable $y = \dot{x}$, we obtain canonical equations for the system in the form

$$\dot{x} = y, \quad \dot{y} = -\frac{2x}{t^2} - \frac{4y}{t} + \frac{u}{t^2}.$$

Putting $u = 0$ and eliminating y, we arrive at the equation

$$t^2\ddot{x} + 4t\dot{x} + 2x = 0.$$

A change of independent variable to s by the transformation $t = e^s$, puts this equation into the form

$$\frac{d^2x}{ds^2} + 3\frac{dx}{ds} + 2x = 0$$

and this has the general solution

$$x = Ae^{-s} + Be^{-2s} = \frac{A}{t} + \frac{B}{t^2}.$$

It then follows that

$$y = \dot{x} = -\frac{A}{t^2} - \frac{2B}{t^3}.$$

The solution satisfying the conditions $x = 1, y = 0$ at $t = \tau$ is now found to be

$$x = \frac{2\tau}{t} - \frac{\tau^2}{t^2}, \quad y = -\frac{2\tau}{t^2} + \frac{2\tau^2}{t^3}.$$

The conditions $x = 0, y = 1$ at $t = \tau$, yield the solution

$$x = \frac{\tau^2}{t} - \frac{\tau^3}{t^2}, \quad y = -\frac{\tau^2}{t^2} + \frac{2\tau^3}{t^3}.$$

The state transition matrix can now be constructed in the form

$$\phi(t, \tau) = \begin{bmatrix} \dfrac{2\tau}{t} - \dfrac{\tau^2}{t^2} & \dfrac{\tau^2}{t} - \dfrac{\tau^3}{t^2} \\ -\dfrac{2\tau}{t^2} + \dfrac{2\tau^2}{t^3} & -\dfrac{\tau^2}{t^2} + \dfrac{2\tau^3}{t^3} \end{bmatrix}.$$

Putting $t^0 = 1$, $x^0 = 0$, $u = 1$, in equation (2.2.20), the response required is

$$\begin{bmatrix} x(t) \\ y(t) \end{bmatrix} = \int_1^t \phi(t, \tau) \begin{bmatrix} 0 \\ 1/\tau^2 \end{bmatrix} d\tau$$

$$= \int_1^t \begin{bmatrix} \dfrac{1}{t} - \dfrac{\tau}{t^2} \\ -\dfrac{1}{t^2} + \dfrac{2\tau}{t^3} \end{bmatrix} d\tau$$

$$= \begin{bmatrix} \dfrac{1}{2} - \dfrac{1}{t} + \dfrac{1}{2t^2} \\ \dfrac{1}{t^2} - \dfrac{1}{t^3} \end{bmatrix}.$$

2.3 Infinite series of matrices

Certain functions of an $M \times M$ matrix A are conveniently defined by means of infinite power series. Such a series will arise in the next section and, in preparation, we shall now establish some of their relevant properties.

Let $F(z)$ be an analytic function of the complex variable z, regular in a neighbourhood of the origin. Then this function can be expanded in a power series, thus,

$$F(z) = a_0 + a_1 z + a_2 z^2 + \cdots, \qquad (2.3.1)$$

convergent for $|z| < R$, R being the radius of the circle of convergence.

The corresponding function $F(A)$ of the matrix A is defined by the equation

$$F(A) = a_0 I + a_1 A + a_2 A^2 + \cdots, \qquad (2.3.2)$$

where the right-hand member is an $M \times M$ matrix each of whose elements is an infinite series. For the definition to be meaningful, each of these series must converge.

To study convergence, we first note that if $A = (a_{ij})$, $B = (b_{ij})$ are two $M \times M$ matrices and if $C = (c_{ij}) = AB$, then

$$|c_{ik}| = |\sum_j a_{ij} b_{jk}| \leqslant \sum_j |a_{ij}| |b_{jk}| \leqslant M\alpha\beta, \qquad (2.3.3)$$

where $\alpha = \max(|a_{ij}|)$, $\beta = \max(|b_{ij}|)$. Repeated application of this result shows that the modulus of the ikth element of A^n is not greater than $M^{n-1}\alpha^n$. Thus, the modulus of the ikth element of $a_n A^n$ is not greater than $|a_n|M^{n-1}\alpha^n$ and it follows that the M^2 infinite series, which are the elements of $F(A)$, are all absolutely convergent (by the comparison test) provided $\sum_n |a_n|M^{n-1}\alpha^n$ is convergent. But this series is known to converge if $M\alpha < R$; this, therefore, provides us with a sufficient condition for the definition (2.3.2) to be meaningful.

As an example, we define e^A or $\exp A$ by the series

$$\exp A = I + A + \frac{1}{2!}A^2 + \frac{1}{3!}A^3 + \cdots. \quad (2.3.4)$$

In this case, R is infinite and the condition $M\alpha < R$ is always satisfied; thus, this definition is valid for all matrices A. Some consequences of this definition will now be derived for later use.

First, provided A and B commute, the identity

$$\exp A \cdot \exp B = \exp(A + B) \quad (2.3.5)$$

can now be established by multiplication of series, exactly as in the elementary case when A, B are scalars; at various points in the argument, it is necessary to rearrange the order of factors in a term; this explains the requirement that $AB = BA$.

Taking $B = -A$, equation (2.3.5) reduces to

$$\exp A \cdot \exp(-A) = I. \quad (2.3.6)$$

This implies that

$$\exp(-A) = [\exp A]^{-1}. \quad (2.3.7)$$

Suppose the matrix A is independent of t. Then

$$\frac{d}{dt}\exp(tA) = \frac{d}{dt}\left[I + tA + \frac{t^2}{2!}A^2 + \frac{t^3}{3!}A^3 + \cdots\right]$$
$$= A + tA^2 + \frac{t^2}{2!}A^3 + \cdots$$
$$= A \exp(tA), \quad (2.3.8)$$

the term-by-term differentiation being valid for any power series within its circle of convergence.

To revert to the general case, provided R is sufficiently large the function $F(A)$ can always be expressed as a polynomial in A of degree $(M - 1)$. This is a consequence of the Cayley-Hamilton theorem, which states that A always satisfies its own characteristic equation. Thus, let $\alpha_1, \alpha_2, \ldots, \alpha_M$ be the eigenvalues of A, assumed distinct. Then

$$|A - \alpha I| = P(\alpha) = (\alpha_1 - \alpha)(\alpha_2 - \alpha) \cdots (\alpha_M - \alpha). \quad (2.3.9)$$

The Cayley-Hamilton theorem asserts that $P(A) \equiv 0$. Assuming $R > \max |\alpha_i|$, so that $F(\alpha_1), F(\alpha_2), \ldots F(\alpha_M)$ all exist, let $S(z)$ be the polynomial of degree $(M - 1)$ such that

$$S(\alpha_i) = F(\alpha_i); \quad (2.3.10)$$

there is just one such polynomial. Then, $F(z) - S(z)$ is regular within the circle $|z| = R$ and has zeros at the points $z = \alpha_i$ ($i = 1, 2, \ldots, M$). It follows that the function $Q(z)$ defined by

$$Q(z) = \frac{F(z) - S(z)}{P(z)} \quad (2.3.11)$$

is also regular within this circle. Thus $F(z)$ can be expressed in the form

$$F(z) = P(z)Q(z) + S(z) \quad (2.3.12)$$

for $|z| < R$. Replacing z by the matrix A in this identity we deduce that

$$F(A) = P(A)Q(A) + S(A) = S(A). \quad (2.3.13)$$

This result can also be extended to the case when the eigenvalues of A are not all distinct.

PROBLEM 9. If

$$A = \pi \begin{bmatrix} 2 & -3 \\ 1 & -2 \end{bmatrix},$$

show that $\cos A = -I$, $\sin A = 0$.

Solution: The eigenvalues of A are $\pm \pi$. Hence, S is a polynomial of the first degree. Taking $S(z) = s_0 + s_1 z$, when $F(z) = \cos z$, equations (2.3.10) require that

$$s_0 + s_1 \pi = \cos \pi = -1,$$

$$s_0 - s_1 \pi = \cos(-\pi) = -1.$$

Hence, $s_0 = -1$, $s_1 = 0$ and $\cos A = S(A) = -I$.

If $F(z) = \sin z$, equations (2.3.10) yield $s_0 = s_1 = 0$. Thus $\sin A = 0$.

2.4 Autonomous linear systems

If the linear system governed by equation (2.2.15) is autonomous, the matrix A is independent of t. In this case, equation (2.2.6) can be solved for the state transition matrix $\phi(t, t^0)$ without difficulty thus:

$\phi(t, t^0)$ is the unique solution of equation (2.2.6) satisfying the initial condition $\phi(t^0, t^0) = I$. But, it follows from the result (2.3.8) that

$$\phi = [\exp(tA)]\phi^0, \qquad (2.4.1)$$

where ϕ^0 is an arbitrary constant $M \times M$ matrix, satisfies this equation. If we choose ϕ^0 so that

$$I = [\exp(t^0 A)]\phi^0, \qquad (2.4.2)$$

the initial condition is also satisfied and equation (2.4.1) must yield the state transition matrix. Thus, we take

$$\phi^0 = [\exp(t^0 A)]^{-1} = \exp(-t^0 A), \qquad (2.4.3)$$

using the identity (2.3.7). Substituting for ϕ^0 in equation (2.4.1), we get

$$\phi(t, t^0) = \exp(tA)\exp(-t^0 A) = \exp[(t - t^0)A]. \qquad (2.4.4)$$

Having expressed the exponential as a polynomial in A by the method explained in the previous section, the response of the system to any control can be found from equation (2.2.20).

PROBLEM 10. A second order linear control system, whose control vector is $u = [u_1, u_2]^T$, is governed by the equation

$$\frac{d^2 y}{dt^2} + 4\frac{dy}{dt} + 3y = \frac{d^2 u_1}{dt^2} + \frac{d^2 u_2}{dt^2} - u_2.$$

Obtain state equations in canonical form and calculate the state transition matrix. Hence determine the response to the control $u(t) = [0, 1]^T$, $t > 0$, if the initial state at $t = 0$ is specified by $y = \dot{y} = 0$, $u = \dot{u} = [0, 0]^T$.

Solution: Write the governing equation in the form

$$\frac{d}{dt}(\dot{y} + 4y - \dot{u}_1 - \dot{u}_2) = -3y - u_2$$

and put
$$x_1 = \dot{y} + 4y - \dot{u}_1 - \dot{u}_2. \quad (2.4.5)$$
Then,
$$\dot{x}_1 = -3y - u_2. \quad (2.4.6)$$

Equation (2.4.5) can be written

$$\frac{d}{dt}(y - u_1 - u_2) = x_1 - 4y. \quad (2.4.7)$$

Putting
$$x_2 = y - u_1 - u_2, \quad (2.4.8)$$
we have
$$\dot{x}_2 = x_1 - 4y. \quad (2.4.9)$$

Using equation (2.4.8) to eliminate y from equations (2.4.6), (2.4.9), we finally obtain the state equations in canonical form, thus:

$$\left.\begin{array}{l}\dot{x}_1 = -3x_2 - 3u_1 - 4u_2, \\ \dot{x}_2 = x_1 - 4x_2 - 4u_1 - 4u_2.\end{array}\right\} \quad (2.4.10)$$

The matrices A, B take the forms

$$A = \begin{bmatrix} 0 & -3 \\ 1 & 4 \end{bmatrix}, \quad B = \begin{bmatrix} -3 & -4 \\ -4 & -4 \end{bmatrix},$$

and A is found to have eigenvalues -1 and -3. The state transition matrix is $\phi(t, \tau) = \exp[(t - \tau)A]$. We know that this function of A can be reduced to a polynomial in A of the first degree. Denoting this by $s_0 I + s_1 A$, the coefficients s_0, s_1 must satisfy the equations

Hence
$$s_0 - s_1 = e^{\tau - t}, \quad s_0 - 3s_1 = e^{3(\tau - t)}.$$

$$s_0 = \tfrac{1}{2}[3e^{\tau - t} - e^{3(\tau - t)}], \quad s_1 = \tfrac{1}{2}[e^{\tau - t} - e^{3(\tau - t)}],$$

and it follows that

$$\phi(t, \tau) = s_0 I + s_1 A$$
$$= \frac{1}{2}\begin{bmatrix} 3e^{\tau - t} - e^{3(\tau - t)} & -3e^{\tau - t} + 3e^{3(\tau - t)} \\ e^{\tau - t} - e^{3(\tau - t)} & -e^{\tau - t} + 3e^{3(\tau - t)} \end{bmatrix}.$$

Substituting the stated initial conditions at $t = 0$ into equations (2.4.5), (2.4.8), we find that $x^0 = [0, 0]^T$. Equation (2.2.20) can now be written down, thus:

$$x(t) = \frac{1}{2}\int_0^t \begin{bmatrix} 3e^{\tau-t} - e^{3(\tau-t)} & -3e^{\tau-t} + 3e^{3(\tau-t)} \\ e^{\tau-t} - e^{3(\tau-t)} & -e^{\tau-t} + 3e^{3(\tau-t)} \end{bmatrix} \begin{bmatrix} -3 & -4 \\ -4 & -4 \end{bmatrix} \begin{bmatrix} 0 \\ 1 \end{bmatrix} d\tau$$

$$= -4\int_0^t \begin{bmatrix} e^{3(\tau-t)} \\ e^{3(\tau-t)} \end{bmatrix} d\tau$$

$$= -\frac{4}{3}\begin{bmatrix} 1 - e^{-3t} \\ 1 - e^{-3t} \end{bmatrix}.$$

Hence

$$x_1 = x_2 = \tfrac{4}{3}(e^{-3t} - 1)$$

determines the response.

Equation (2.4.8) shows that the y-response is given by

$$y = \tfrac{1}{3}(4e^{-3t} - 1).$$

(N.B. y is discontinuous at $t = 0$; this discontinuity is caused by our assumption that the value of u_2 changes instantaneously from 0 to 1 at this instant. It is an important feature of the state equations in canonical form that finite discontinuities in the control variables do not lead to discontinuities in the state variables, but only in their derivatives.) ●

Returning to the general case of the autonomous linear system governed by equation (2.2.15), it follows from equation (2.3.10) that the polynomial $S(A)$ which is the state transition matrix has coefficients determined by the M equations

$$S(\alpha_i) = \exp[(t - \tau)\alpha_i], \qquad (2.4.11)$$

$i = 1, 2, \ldots, M$, the α_i being the eigenvalues of the matrix A. It follows that these coefficients will be linearly dependent upon the quantities $\exp[(t - \tau)\alpha_i]$ and, thus, will tend to zero as $t \to +\infty$ only if the real parts of all the eigenvalues α_i are negative. This, then, is the condition for $\phi(t, \tau) = S(A)$ to tend to zero as $t \to +\infty$.

If this condition is satisfied and the system is permitted to run freely without control from an arbitrary initial state x^0, since $x(t) = \phi(t, t^0)x^0$, it follows that $x(t) \to 0$ as $t \to +\infty$. In these circumstances, if the system is disturbed from the equilibrium state $x = 0$ at an initial instant, it will settle back into this state after some time has elapsed. A system having this property is said to be *stable*. If, however, any of the real parts of the eigenvalues α_i are greater than zero, some or all of the components of $x(t)$ will increase in magnitude without limit as $t \to +\infty$; the system is then said to be *unstable*.

2.5 Optimal control

Returning now to the general non-linear system whose state equations have been given at (2.1.1), we shall suppose that the control vector function $u(t)$ is to be chosen in such a way that a specified performance index is to be maximized or a certain cost is to be minimized. In either case, we shall adopt the notation of Chapter 1 and denote the quantity to be optimized by C. We shall suppose that the control operates over a fixed time interval $t^0 \leqslant t \leqslant t^1$ and that, when the response has been found, C is to be calculated from a formula of the type

$$C = G(x^1) + \int_{t^0}^{t^1} g(x, u, t)\, dt, \qquad (2.5.1)$$

where x^1 denotes the state vector at the final instant $t = t^1$. Thus, C is taken to depend partly on the final state (which, for the present, is supposed subject to no constraints) and partly on the behaviour of the system in reaching this final state. In particular cases, we may take g to be identically zero, in which case the performance of the system is being assessed by reference to the final state achieved alone. If it is desired to give no special weight to the final state, then G should be put to zero. Dependence of G upon the final control vector u^1 has been excluded, for the reason that such dependence would generalize the problem to a trivial degree only. This arises from the fact that we shall permit $u(t)$ to be discontinuous and, hence, its final value can then be chosen independently of its earlier values; consequently if G were made dependent upon u^1, maximization or minimization of G with respect to u^1 would simply have the effect of introducing an isolated discontinuity of $u(t)$ at this terminal. Such a discontinuity will rarely be of practical significance.

The functions g, G will be assumed to possess continuous second order derivatives with respect to all their arguments over a sufficiently extensive open region O of the xut-space containing the optimal trajectories.

2.6 Necessary conditions for optimal control

The problem to be solved can be stated as follows: Supposing the system to be in a known initial state $x = x(t^0) = x^0$ at $t = t^0$ it is required to determine a continuous vector control function $u(t)$ over a fixed interval $[t^0, t^1]$, such that C is minimized. Later, $u(t)$ will be permitted to have a finite number of finite discontinuities, but this complication is declared inadmissible for the present. It will be assumed that such an optimal control exists and it will, as usual, be denoted by $u(t)$; $x(t)$ will then describe the associated optimal behaviour. To establish necessary conditions to be satisfied by u and x, we consider a neighbouring non-optimal control determined by $v(t, \epsilon) = u(t) + \epsilon \xi(t)$, where $\xi(t)$ is an arbitrary continuous vector function over $[t^0, t^1]$ and ϵ is a positive or negative parameter sufficiently small to ensure that the associated trajectories lie within the open region O referred to above. The system's non-optimal behaviour will be described by a state vector function $y(t, \epsilon)$; since the initial state is fixed, we must have

$$y(t^0, \epsilon) = x^0 \qquad (2.6.1)$$

identically in ϵ.

Clearly,

$$v(t, 0) = u(t), \qquad y(t, 0) = x(t). \qquad (2.6.2)$$

Hence, by this means, the optimal trajectory has been exhibited as a member of a family of trajectories, each of which satisfies the differential constraints (2.1.2) and the initial conditions. We shall refer to such a family as a family of *admissible trajectories* and to the mathematical procedure whereby the family is constructed about the optimal trajectory as an *embedding* of the optimal trajectory within the family.

It is now possible to prove an embedding theorem, relating to the continuity and differentiability properties of the vector function $y(t, \epsilon)$. The reader who is interested in the details should refer to the standard work by G. A. Bliss (bibliography, item 1). It will here simply be assumed that $y(t, \epsilon)$ possesses continuous first derivatives with respect to both its variables in the region $t^0 \leqslant t \leqslant t^1$, $|\epsilon| < \delta$.

We note that

$$v_r = u_r + \epsilon \xi_r, \qquad \frac{\partial v_r}{\partial \epsilon} = \xi_r. \qquad (2.6.3)$$

ξ_r is termed the *variation* of u_r (some texts call $\epsilon \xi_r$ the variation of

u_r). The corresponding equations for the state vector will be taken to be

$$y_i = y_i(t, \epsilon), \left(\frac{\partial y_i}{\partial \epsilon}\right)_{\epsilon=0} = \eta_i(t). \quad (2.6.4)$$

If ϵ is small, the equation

$$y_i = y_i(t, 0) + \epsilon\left(\frac{\partial y_i}{\partial \epsilon}\right)_{\epsilon=0} = x_i + \epsilon\eta_i \quad (2.6.5)$$

is approximately true (to first order in ϵ). We shall accordingly refer to η_i as the variation of x_i (again, some authors take $\epsilon\eta_i$ to be this variation). Differentiating equation (2.6.1) with respect to ϵ and putting $\epsilon = 0$, it is seen that the η_i satisfy the initial condition

$$\eta_i(t^0) = \eta_i^0 = 0. \quad (2.6.6)$$

Since $y(t, \epsilon)$ describes a possible behaviour of the system, it must satisfy the state equation (2.1.2) for all values of ϵ; thus,

$$\frac{\partial y}{\partial t} = f(y, v, t). \quad (2.6.7)$$

Differentiating the ith equation of this set with respect to ϵ, we get

$$\frac{\partial^2 y_i}{\partial \epsilon \, \partial t} = \frac{\partial f_i}{\partial y_j}\frac{\partial y_j}{\partial \epsilon} + \frac{\partial f_i}{\partial v_r}\frac{\partial v_r}{\partial \epsilon}. \quad (2.6.8)$$

Putting $\epsilon = 0$, this equation can be written

$$\dot{\eta}_i = \frac{\partial f_i}{\partial x_j}\eta_j + \frac{\partial f_i}{\partial u_r}\xi_r. \quad (2.6.9)$$

These are called the *equations of variation* along the optimal trajectory. Assuming this trajectory to be known and the variation $\xi_r(t)$ to be given, these provide a set of linear equations for the variations $\eta_i(t)$ which, together with the initial conditions (2.6.6), determine them uniquely.

Employing equation (2.5.1), the cost associated with the control v is given by

$$C = C(\epsilon) = G(y^1) + \int_{t^0}^{t^1} g(y, v, t)\, dt. \quad (2.6.10)$$

This expresses C as a function of the parameter ϵ and leads to a necessary condition for optimality, namely $dC/d\epsilon = 0$ at $\epsilon = 0$.

Now

$$\frac{dC}{d\epsilon} = \frac{\partial G}{\partial y_i^1}\frac{dy_i^1}{d\epsilon} + \int_{t^0}^{t^1}\left(\frac{\partial g}{\partial y_i}\frac{\partial y_i}{\partial \epsilon} + \frac{\partial g}{\partial v_r}\frac{\partial v_r}{\partial \epsilon}\right)dt. \quad (2.6.11)$$

Thus, putting $\epsilon = 0$, the necessary condition becomes

$$\frac{\partial G}{\partial x_i^1}\eta_i^1 + \int_{t^0}^{t^1}\left(\frac{\partial g}{\partial x_i}\eta_i + \frac{\partial g}{\partial u_r}\xi_r\right)dt = 0. \quad (2.6.12)$$

To bring this condition into a more tractable form, it is helpful to introduce a set of *multipliers* $\lambda_i(t)$ satisfying the *adjoint system of equations*

$$\dot{\lambda}_i = -\lambda_j\frac{\partial f_j}{\partial x_i} - \frac{\partial g}{\partial x_i}. \quad (2.6.13)$$

These equations, together with the end conditions

$$\lambda_i(t^1) = \lambda_i^1 = \frac{\partial G}{\partial x_i^1}, \quad (2.6.14)$$

uniquely determine the multipliers along the optimal trajectory. Then,

$$\frac{d}{dt}(\lambda_i\eta_i) = \dot{\lambda}_i\eta_i + \lambda_i\dot{\eta}_i = \lambda_i\frac{\partial f_i}{\partial u_r}\xi_r - \eta_i\frac{\partial g}{\partial x_i}, \quad (2.6.15)$$

having employed equations (2.6.9) and (2.6.13). It follows that the condition (2.6.12) is equivalent to

$$\lambda_i^1\eta_i^1 + \int_{t^0}^{t^1}\left[\left(\frac{\partial g}{\partial u_r} + \lambda_i\frac{\partial f_i}{\partial u_r}\right)\xi_r - \frac{d}{dt}(\lambda_i\eta_i)\right]dt = 0. \quad (2.6.16)$$

We now introduce the *Hamiltonian*

$$H = g + \lambda_i f_i. \quad (2.6.17)$$

Since

$$\int_{t^0}^{t^1}\frac{d}{dt}(\lambda_i\eta_i)\,dt = \lambda_i^1\eta_i^1 - \lambda_i^0\eta_i^0 \quad (2.6.18)$$

and $\eta_i^0 = 0$ (equation (2.6.6)), the condition (2.6.16) finally reduces to

$$\int_{t^0}^{t^1}\frac{\partial H}{\partial u_r}\xi_r\,dt = 0. \quad (2.6.19)$$

But, this condition must be satisfied for arbitrary continuous functions $\xi_r(t)$. We therefore conclude that

$$\frac{\partial H}{\partial u_r} = 0 \qquad (2.6.20)$$

along an optimal trajectory.

It should be observed that our construction of H allows the adjoint equations (2.6.13) to be written in an easily remembered form, namely

$$\dot{\lambda}_i = -\frac{\partial H}{\partial x_i}. \qquad (2.6.21)$$

Further, the state equations (2.1.1) can be expressed in the form

$$\dot{x}_i = \frac{\partial H}{\partial \lambda_i}. \qquad (2.6.22)$$

The two sets of equations (2.6.21), (2.6.22) will be referred to as *Hamilton's equations* for this optimal control problem. These equations first arose, of course, in Hamilton's development of the theory of mechanical systems.

Equations (2.6.20)–(2.6.22) provide $(2M + N)$ equations for the same number of unknown functions $x_i(t)$, $\lambda_i(t)$, $u_r(t)$, and, together with the initial conditions $x = x^0$ and the end conditions (2.6.14), are enough to determine the optimal behaviour. However, these equations do not constitute sufficient conditions for optimality and further investigation is therefore needed to establish whether or not the behaviour so calculated is truly optimal. Very often the nature of the presumed optimal behaviour is such that its acceptance as such without further analysis can be considered reasonable and we shall very often take this step in particular problems below. However, we shall return to the study of this matter in section 5.5.

PROBLEM 11. At $t = 0$, the capacitor C in the network shown in Fig. 2.1 is charged so that the potential difference across its plates is x^0. $u(t)$ is a potential difference applied across the terminals A, B and $x(t)$ is the resulting potential difference across the plates of the capacitor at time $t > 0$. Calculate $u(t)$ for $0 \leqslant t \leqslant 1$, if

$$\int_0^1 (x^2 + \tfrac{1}{5}u^2)\, dt$$

is to be minimized. Assume $C = 1/R$.

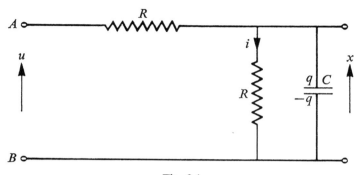

Fig. 2.1

Solution: Observe that the form taken by the cost implies that our control is required to reduce the magnitude of the potential difference between the plates of C as soon as possible but, at the same time, the control variable u must not take large values. If the term $u^2/5$ were omitted from the integrand of the cost, the whole emphasis would be on reducing the magnitude of x without consideration of the values taken by u. In these circumstances, the problem would have no solution, since letting $u \to -\infty$ initially would reduce x to zero.

At time t, let i be the current in the resistor shunted across the capacitor and let q, $-q$ be the charges on its plates. Then, the current through the other resistor is $i + dq/dt$ and, hence,

$$u = R(i + dq/dt) + x.$$

Since $x = Ri = q/C = Rq$, elimination of i and q leads to the state equation

$$\dot{x} = -2x + u. \tag{2.6.23}$$

The Hamiltonian for the problem is

$$H = x^2 + \tfrac{1}{5}u^2 + \lambda(-2x + u).$$

Conditions (2.6.20), (2.6.21) then yield

$$\tfrac{2}{5}u + \lambda = 0, \qquad \dot{\lambda} = -2x + 2\lambda. \tag{2.6.24}$$

We now have three equations for u, x and λ. Eliminating λ between equations (2.6.24), we obtain

$$\dot{u} - 2u = 5x. \tag{2.6.25}$$

This is the *optimal control law*. It is possible to design an electronic device which accepts x as input, generates u as output and which is governed by the last equation. If this device is coupled between the capacitor and the terminals AB, it will ensure that the system's behaviour is always optimal from any initial state.

Solving equations (2.6.23), (2.6.25) for x and u, we get a general solution

$$x = Ae^{3t} + Be^{-3t}, \qquad u = 5Ae^{3t} - Be^{-3t}.$$

The constants A, B can now be determined from the initial condition $x = x^0$ at $t = 0$ and the terminal condition (2.6.14), i.e. $\lambda = 0$ (and, hence, $u = 0$) at $t = 1$. The results are

$$x = x^0 \frac{5e^{3(1-t)} + e^{-3(1-t)}}{5e^3 + e^{-3}},$$

$$u = -x^0 \frac{5(e^{3(1-t)} - e^{-3(1-t)})}{5e^3 + e^{-3}}.$$

These equations specify the optimal behaviour and control. The minimal cost can be calculated by substitution for x and u in the cost integral. We find that

$$C_{\min} = \frac{e^3 - e^{-3}}{5e^3 + e^{-3}} (x^0)^2. \qquad \bullet$$

PROBLEM 12. The thrust generated by a rocket's motor is a known function $P(t)$ of the time from ignition $t = 0$ to "all-burnt" $t = T$. Calculate the optimal thrust direction programme if the rocket's range over the horizontal plane through the launching point is to be maximized. Neglect air resistance, curvature and rotation of the earth and variation of gravity.

Solution: Ox, Oy are horizontal and vertical axes through the launching point O (Fig. 2.2). We shall assume that the rocket moves in the planes of these axes and that the coordinates of its centre of mass at time t are (x, y). Then, if M is the rocket mass and θ is the angle made by the direction of thrust with the horizontal, equations of motion for the vehicle are

$$M\ddot{x} = P \cos \theta, \qquad M\ddot{y} = P \sin \theta - Mg$$

(g is the gravitational acceleration). Putting $u = \dot{x}$, $v = \dot{y}$, $a = P/M$, these equations are easily seen to be equivalent to the canonical

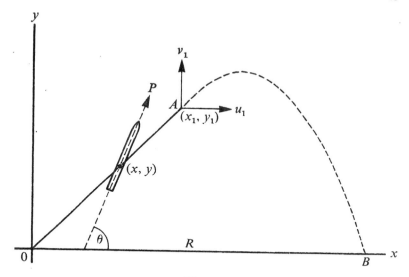

Fig. 2.2

system
$$\dot{x} = u, \qquad \dot{y} = v,$$
$$\dot{u} = a\cos\theta, \qquad \dot{v} = a\sin\theta - g. \qquad (2.6.26)$$

In these equations, x, y, u, v are state variables and θ is the control variable; $a(t)$ is assumed to be a known function.

At "all-burnt" (A in Fig. 2.1), $t = T$, $x = x_1$, $y = y_1$, $u = u_1$, $v = v_1$. Thereafter, the rocket falls freely under gravity, describing a parabolic arc which intersects the x-axis at B. The range of the rocket is $OB = R$ and this is given by

$$R = x_1 + \frac{u_1}{g}[v_1 + \sqrt{(v_1^2 + 2gy_1)}].$$

$\theta(t)$ is to be chosen so that R is maximized.

The Hamiltonian for the problem is

$$H = \lambda_x u + \lambda_y v + \lambda_u a\cos\theta + \lambda_v(a\sin\theta - g).$$

Optimal control is determined by the equation

$$\frac{\partial H}{\partial \theta} = -\lambda_u a \sin\theta + \lambda_v a \cos\theta = 0,$$

i.e.
$$\tan\theta = \lambda_v/\lambda_u. \qquad (2.6.27)$$

Hamilton's equations are

$$\begin{aligned}\dot\lambda_x = \dot\lambda_y = 0,\\ \dot\lambda_u = -\lambda_x, \quad \dot\lambda_v = -\lambda_y,\end{aligned} \qquad (2.6.28)$$

and their general solution is

$$\lambda_x = A, \quad \lambda_y = B, \quad \lambda_u = C - At, \quad \lambda_v = D - Bt,$$

where A, B, C, D are constants.

The terminal conditions (2.6.14) have to be applied at $t = T$; they are

$$\lambda_x = \frac{\partial R}{\partial x_1} = 1, \qquad \lambda_y = \frac{\partial R}{\partial y_1} = u_1/r,$$

$$\lambda_u = \frac{\partial R}{\partial u_1} = (v_1 + r)/g, \qquad \lambda_v = \frac{\partial R}{\partial v_1} = \frac{u_1}{g}\left(1 + \frac{v_1}{r}\right),$$

where $r = \sqrt{(v_1^2 + 2gy_1)}$. These conditions fix A, B, C, D and the multipliers are then given at any time t by

$$\lambda_x = 1, \quad \lambda_y = u_1/r,$$
$$\lambda_u = T - t + (v_1 + r)/g,$$
$$\lambda_v = \frac{u_1}{r}[T - t + (v_1 + r)/g].$$

It now follows from the optimal control equation (2.6.27) that

$$\tan\theta = u_1/\sqrt{(v_1^2 + 2gy_1)}. \qquad (2.6.29)$$

This shows that the thrust direction must be maintained constant.

The terminal values x_1, y_1, u_1, v_1 have still to be calculated. Knowing that θ is constant, the equations of motion (2.6.26) can be integrated over the time interval $[0, T]$ from the initial conditions $x = y = u = v = 0$. Before this can be done, the acceleration $a(t)$ must be specified as a function of t. In the simple special case when

a is constant, we easily calculate that

$$x_1 = \tfrac{1}{2}aT^2 \cos \theta, \qquad y_1 = \tfrac{1}{2}(a \sin \theta - g)T^2,$$
$$u_1 = aT \cos \theta, \qquad v_1 = (a \sin \theta - g)T.$$

Substituting from these equations into equation (2.6.29), the reader may verify that the following equation for θ can be derived

$$\sin^3 \theta - 2n \sin^2 \theta + n = 0,$$

where $n = a/g$. Provided $n > 1$, this equation has a single root for $\sin \theta$ lying in the range $0 < \sin \theta < 1$. ●

PROBLEM 13. An artificial satellite is rotating about its centre of mass. Relative to principal axes of inertia, the components of its angular momentum are x_i. If u_i are the components of the controlling torque, the equations of motion are

$$\left. \begin{aligned} \dot{x}_1 &= \alpha_1 x_2 x_3 + u_1, \\ \dot{x}_2 &= \alpha_2 x_3 x_1 + u_2, \\ \dot{x}_3 &= \alpha_3 x_1 x_2 + u_3, \end{aligned} \right\} \qquad (2.6.30)$$

where $\alpha_1 + \alpha_2 + \alpha_3 = 0$. If the u_i are to be chosen so that, from a given initial state of rotation, the cost

$$C = \frac{1}{2} \int_0^\infty [q(x_1^2 + x_2^2 + x_3^2) + \frac{1}{q}(u_1^2 + u_2^2 + u_3^2)] \, dt$$

is to be minimized, verify that the control law $u_i = -qx_i$ satisfies the optimality conditions and that this control ultimately reduces the angular momentum to zero.*

Solution: The Hamiltonian for the problem is

$$H = \tfrac{1}{2}q(x_1^2 + x_2^2 + x_3^2) + \frac{1}{2q}(u_1^2 + u_2^2 + u_3^2)$$
$$+ \lambda_1(\alpha_1 x_2 x_3 + u_1) + \lambda_2(\alpha_2 x_3 x_1 + u_2) + \lambda_3(\alpha_3 x_1 x_2 + u_3).$$

The vanishing of the derivatives $\partial H/\partial u_i$ leads to the equations

$$u_i = -q\lambda_i, \qquad i = 1, 2, 3. \qquad (2.6.31)$$

* See: Debs, A. S., Athans, M., On the optimal angular velocity control of asymmetrical space vehicles, *IEEE Trans. Auto. Cont.*, 1969, 80–83.

Hamilton's equations are

$$\dot\lambda_1 = -\frac{\partial H}{\partial x_1} = -qx_1 - \lambda_2\alpha_2 x_3 - \lambda_3\alpha_3 x_2.$$

$$\dot\lambda_2 = -\frac{\partial H}{\partial x_2} = -qx_2 - \lambda_3\alpha_3 x_1 - \lambda_1\alpha_1 x_3,$$

$$\dot\lambda_3 = -\frac{\partial H}{\partial x_3} = -qx_3 - \lambda_1\alpha_1 x_2 - \lambda_2\alpha_2 x_1.$$

Eliminating the λ_i by use of equation (2.6.31), these become

$$\begin{aligned}\dot u_1 &= q^2 x_1 - \alpha_2 u_2 x_3 - \alpha_3 u_3 x_2,\\ \dot u_2 &= q^2 x_2 - \alpha_3 u_3 x_1 - \alpha_1 u_1 x_3,\\ \dot u_3 &= q^2 x_3 - \alpha_1 u_1 x_2 - \alpha_2 u_2 x_1.\end{aligned} \quad (2.6.32)$$

The terminal conditions (2.6.14) require the λ_i to vanish at the final instant. Hence, $u_i \to 0$ as $t \to \infty$.

All these conditions can be satisfied by taking the control law to be $u_i = -qx_i$. For then, since $\alpha_1 + \alpha_2 + \alpha_3 = 0$, the two systems of equations (2.6.30), (2.6.32) both reduce to the system

$$\begin{aligned}\dot x_1 &= \alpha_1 x_2 x_3 - qx_1,\\ \dot x_2 &= \alpha_2 x_3 x_1 - qx_2,\\ \dot x_3 &= \alpha_3 x_1 x_2 - qx_3,\end{aligned}$$

and any set of initial conditions $x = x^0$, $t = 0$ determines a unique solution of these equations. Further, this solution will be such that $x_i \to 0$ as $t \to \infty$; for, multiplying the equations by x_1, x_2, x_3 successively and adding, we get

$$\frac{d}{dt}[\tfrac{1}{2}(x_1^2 + x_2^2 + x_3^2)] = -q(x_1^2 + x_2^2 + x_3^2);$$

thus,

$$x_1^2 + x_2^2 + x_3^2 = (x_1^2 + x_2^2 + x_3^2)^0 e^{-2qt} \quad (2.6.33)$$

and, hence,

$$x_1^2 + x_2^2 + x_3^2 \to 0 \quad \text{as} \quad t \to \infty.$$

It follows that $x_i \to 0$, and, since $u_i = -qx_i$, $u_i \to 0$ also.

Thus, all the optimality conditions are satisfied by this control law and it is reasonable to assume it results in C being minimized.

Making use of equation (2.6.33), we find that

$$C_{\min} = \tfrac{1}{2}(x_1{}^2 + x_2{}^2 + x_3{}^2)^0.$$

2.7 Optimization of linear systems

Consider a linear control system whose state equations are given at (2.2.1) (with a_{ij}, b_{ir} as functions of t, in general). If we assume that the cost takes the quadratic form below,

$$C = \tfrac{1}{2}s_{ij}x_i^1 x_j^1 + \frac{1}{2}\int_{t^0}^{t^1}(p_{ij}x_ix_j + 2q_{ir}x_iu_r + r_{rs}u_ru_s)\,dt, \quad (2.7.1)$$

where the s_{ij} are constants, but the p_{ij}, q_{ir}, r_{rs}, may be functions of t, a complete optimization theory can be developed. This problem is a generalization of the one for a static system solved in Chapter 1 as Problem 4. Also, a particular case is provided by Problem 11.

We shall assume, without loss of generality, that the matrices $S = (s_{ij})$, $P = (p_{ij})$, $R = (r_{rs})$, are symmetric, i.e. $s_{ij} = s_{ji}$, etc.

The Hamiltonian for the problem is given by

$$H = \tfrac{1}{2}p_{ij}x_ix_j + q_{ir}x_iu_r + \tfrac{1}{2}r_{rs}u_ru_s + \lambda_i(a_{ij}x_j + b_{ir}u_r). \quad (2.7.2)$$

Thus, the control equations take the form

$$\frac{\partial H}{\partial u_r} = q_{ir}x_i + r_{rs}u_s + \lambda_i b_{ir} = 0, \quad (2.7.3)$$

$r = 1, 2, \ldots, N$. Hamilton's equations are:

$$\dot{x}_i = \frac{\partial H}{\partial \lambda_i} = a_{ij}x_j + b_{ir}u_r, \quad (2.7.4)$$

$$\dot{\lambda}_i = -\frac{\partial H}{\partial x_i} = -p_{ij}x_j - q_{ir}u_r - \lambda_j a_{ji}. \quad (2.7.5)$$

The set of equations (2.7.3)–(2.7.5) are conveniently expressed in matrix form, thus:

$$Q^T x + Ru + B^T \lambda = 0, \quad (2.7.6)$$

$$\dot{x} = Ax + Bu, \quad (2.7.7)$$

$$\dot{\lambda} = -Px - Qu - A^T\lambda. \quad (2.7.8)$$

Assuming that R^{-1} exists, equation (2.7.6) can be solved for u, to yield
$$u = -R^{-1}Q^T x - R^{-1}B^T \lambda. \tag{2.7.9}$$
Substituting for u in equations (2.7.7), (2.7.8), we arrive at the system
$$\left.\begin{aligned} \dot{x} &= (A - BR^{-1}Q^T)x - BR^{-1}B^T\lambda, \\ \dot{\lambda} &= (QR^{-1}Q^T - P)x + (QR^{-1}B^T - A^T)\lambda. \end{aligned}\right\} \tag{2.7.10}$$
This system comprises $2M$ linear equations in the $2M$ unknowns x_i, λ_i and is to be solved under the $2M$ end conditions
$$x = x^0 \quad \text{at} \quad t = t^0, \tag{2.7.11}$$
$$\lambda = \lambda^1 = Sx^1 \quad \text{at} \quad t = t^1; \tag{2.7.12}$$
the conditions (2.7.12) arise from equations (2.6.14).

The set of equations (2.7.10) can be solved by the method explained in section 2.2. Since the equations are homogeneous, if $\psi(t, \tau)$ is the $2M \times 2M$ state transition matrix for the system, equation (2.2.8) (with t^0 replaced by t^1) shows that we have
$$\begin{bmatrix} x(t) \\ \lambda(t) \end{bmatrix} = \psi(t, t^1) \begin{bmatrix} x^1 \\ \lambda^1 \end{bmatrix}. \tag{2.7.13}$$
If we now partition ψ into four $M \times M$ matrices ϕ_i ($i = 1, 2, 3, 4$), this last equation can be written in the form
$$\begin{bmatrix} x(t) \\ \lambda(t) \end{bmatrix} = \begin{bmatrix} \phi_1(t, t^1) & \phi_2(t, t^1) \\ \phi_3(t, t^1) & \phi_4(t, t^1) \end{bmatrix} \begin{bmatrix} x^1 \\ \lambda^1 \end{bmatrix} \tag{2.7.14}$$
which, upon expansion yields the following pair of matrix equations
$$\left.\begin{aligned} x(t) &= \phi_1(t, t^1)x^1 + \phi_2(t, t^1)\lambda^1, \\ \lambda(t) &= \phi_3(t, t^1)x^1 + \phi_4(t, t^1)\lambda^1. \end{aligned}\right\} \tag{2.7.15}$$
But, $\lambda^1 = Sx^1$ (equation (2.7.12)) and, hence,
$$x(t) = \theta_1(t, t^1)x^1, \qquad \lambda(t) = \theta_2(t, t^1)x^1, \tag{2.7.16}$$
where
$$\theta_1 = \phi_1 + \phi_2 S, \qquad \theta_2 = \phi_3 + \phi_4 S. \tag{2.7.17}$$
Assuming θ_1^{-1} exists, the equations (2.7.16) imply that
$$\lambda(t) = \theta_2(t, t^1)\theta_1^{-1}(t, t^1)x(t) = K(t, t^1)x(t), \tag{2.7.18}$$
where $K = \theta_2 \theta_1^{-1}$.

Substituting $\lambda = Kx$ in equations (2.7.10), we obtain the equations

$$\dot{x} = (A - BR^{-1}Q^T - BR^{-1}B^TK)x, \qquad (2.7.19)$$

$$\dot{K}x + K\dot{x} = [QR^{-1}Q^T - P + (QR^{-1}B^T - A^T)K]x. \qquad (2.7.20)$$

Substitution for $\dot{x}$ in the second of these equations from the first now yields the result

$$(\dot{K} - KBR^{-1}B^TK + KA + A^TK - KBR^{-1}Q^T - QR^{-1}B^TK$$
$$+ P - QR^{-1}Q^T)x = 0. \qquad (2.7.21)$$

Now, this equation has been derived without any appeal being made to the initial condition (2.7.11). It follows that it is valid for arbitrary x^0 and, hence, for arbitrary x. We conclude that $K(t, t^1)$ must satisfy the equation

$$\dot{K} = (KB + Q)R^{-1}(B^TK + Q^T) - KA - A^TK - P. \qquad (2.7.22)$$

This is a set of first order ordinary differential equations in the elements of K. Since its right-hand member contains terms of the second degree in the dependent variables, it is a *generalized Riccati equation*.

Before this system of equations can be integrated, an end condition must be found. Since $\psi(t^1, t^1) = I$, it follows that $\phi_1(t^1, t^1) = \phi_4(t^1, t^1) = I$ and $\phi_2(t^1, t^1) = \phi_3(t^1, t^1) = 0$. Hence $\theta_1(t^1, t^1) = I$, $\theta_2(t^1, t^1) = S$ and these imply that

$$K(t^1, t^1) = S. \qquad (2.7.23)$$

This provides the required end condition.

Taking the transpose of both members of equation (2.7.22), we get

$$\dot{K}^T = (K^TB + Q)R^{-1}(B^TK^T + Q^T) - A^TK^T - K^TA - P, \qquad (2.7.24)$$

remembering that P and R are symmetric. Thus K^T satisfies exactly the same Riccati equation as K. Further, transposing the end condition (2.7.23), we find that $K^T(t^1, t^1) = S^T = S$, i.e. K^T satisfies the same end condition as K. It now follows that

$$K^T = K \qquad (2.7.25)$$

for all values of t; thus, K is symmetric.

When K has been calculated, the optimal behaviour can be found by integrating equation (2.7.19) for x, employing the initial condition $x = x^0$ at $t = t^0$ to provide starting values. Putting $\lambda = Kx$ in equation (2.7.9), the control law is obtained, namely

$$u = -R^{-1}(Q^T + B^T K)x. \qquad (2.7.26)$$

The superiority for computational purposes of the Riccati equation (2.7.22) over the system of Hamilton equations from which it is derived, lies in the circumstance that the unknown K is subject to end conditions (2.7.23) which are all applicable at the same instant t^1. Thus, if a numerical integration procedure has to be employed, this can commence at $t = t^1$ without difficulty and values of $K(t, t^1)$ can be generated by a straightforward backward integration process. In the case of the system of Hamilton equations, however, x is subject to conditions at $t = t^0$, whereas λ is subject to conditions at $t = t^1$; thus from whichever instant integration commences, values of half of the variables will be unknown and will therefore have to be guessed and subjected to later correction.

The number of unknown elements of K is $\frac{1}{2}M(M + 1)$, whereas the number of dependent variables in Hamilton's equations is $2M$. Hence, unless $M \leq 3$, the order of the Riccati system will exceed that of the Hamilton system. Further, the Riccati equations are non-linear and are accordingly less tractable for solution by analytical methods. For these reasons, the Hamilton equations are usually to be preferred if an analytical solution is to be attempted.

PROBLEM 14. A system has state equation $\dot{x} = u$ and the cost to be minimized is given by

$$C = \tfrac{1}{2}x_1^2 + \tfrac{1}{2}\alpha \int_0^T u^2 \, dt,$$

where $x_1 = x(T)$. Calculate the optimal control from a given initial state $x = x_0$.

Solution: A physical realization of this problem can be given in terms of a rocket vehicle which is constrained to move along a straight line under zero resistance and gravity. Taking x to be its velocity and m to be its mass and assuming that the motor operates at constant power P, it can be proved that

$$\frac{d}{dt}\left(\frac{1}{m}\right) = \frac{1}{2P}\dot{x}^2 = \frac{1}{2P}u^2.$$

CONTROL SYSTEMS

Integration over the interval $(0, T)$ accordingly leads to the result

$$\frac{1}{m_1} - \frac{1}{m_0} = \frac{1}{2P} \int_0^T u^2 \, dt,$$

indicating that the integral part of C can be interpreted as a measure of the propellent consumed.

Minimization of C therefore corresponds to a manoeuvre in which we attempt to reduce the vehicle's velocity from an initial value x_0 to as small a final value x_1 as possible and, at the same time, to prevent the propellent consumption from becoming large. The relative weight we give to these two conflicting factors is determined by the value of the parameter α.

All the matrices which arose in the general theory developed earlier in the section reduce to scalar quantities in this special case. Thus, $A = 0$, $B = 1$, $P = Q = 0$, $R = \alpha$, $S = 1$ and equation (2.7.22) takes the simple form

$$\dot{K} = \frac{1}{\alpha} K^2.$$

Hence, $K = -\alpha/(t + \beta)$, where β is a constant of integration. The condition (2.7.23) requires that $K = 1$ at $t = T$; it follows that

$$K(t, T) = \frac{\alpha}{\alpha + T - t}.$$

The optimal control law is now found from equation (2.7.26) in the form

$$u = -\frac{x}{\alpha + T - t}.$$

Substitution in the state equation yields

$$\dot{x} = -\frac{x}{\alpha + T - t}.$$

Solving this equation under the initial condition $x = x_0$ at $t = 0$, it will be found that the optimal behaviour is specified by the equation

$$x = x_0 \left(1 - \frac{t}{\alpha + T}\right).$$

The optimal control is then given by

$$u = -\frac{x_0}{\alpha + T},$$

i.e. u is constant.

It has been proved, therefore, that optimal control of the rocket described earlier requires regulation of the motor thrust to generate constant retardation $x_0/(\alpha + T)$, thereby reducing the velocity from x_0 to $x_1 = \alpha x_0/(\alpha + T)$ in the available time T. The associated minimal cost is then

$$C_{\min} = \frac{\alpha x_0^2}{2(\alpha + T)}.$$

In the extreme case $\alpha = 0$, no consideration is to be given to the propellent expenditure and the final velocity is made zero by applying a retardation x_0/T.

In the other extreme case $\alpha = \infty$, the propellent expenditure becomes all important. Propellent expenditure is then completely avoided and the optimal retardation is zero. ●

The special case when t^1 is infinite is of some interest. Clearly, if the integral part of the cost function is to converge, it is necessary that both x and u tend to zero as $t \to \infty$. Hence, $x_i^1 = 0$ and we shall take

$$C = \frac{1}{2}\int_{t^0}^{\infty} (p_{ij}x_ix_j + 2q_{ir}x_iu_r + r_{rs}u_ru_s)\, dt. \quad (2.7.27)$$

Before letting $t^1 \to \infty$, it is convenient to put $\tau = t^1 - t$. Then, the matrix K can be regarded as a function of τ satisfying the Riccati equation

$$\frac{dK}{d\tau} = P + KA + A^TK - (KB + Q)R^{-1}(B^TK + Q^T) \quad (2.7.28)$$

and the end condition $K = 0$ at $\tau = 0$. As $t^1 \to \infty$, t remaining finite, $\tau \to \infty$ and we shall assume that $dK/d\tau \to 0$ (if $dK/d\tau$ tends to a non-zero value, $K(\tau)$ must become infinite). Then $K(\tau) \to K_s$, where K_s is a constant symmetric matrix satisfying the *steady state Riccati equation*, namely

$$P + K_sA + A^TK_s - (K_sB + Q)R^{-1}(B^TK_s + Q^T) = 0. \quad (2.7.29)$$

The optimal control now follows as previously.

Being quadratic in K_s, equation (2.7.29) will possess more than one root. However, since the optimal behaviour of the system is determined by equation (2.7.19), if x is not to tend to infinity with t, it is necessary that the matrix

$$A - BR^{-1}(Q^T + B^T K) \qquad (2.7.30)$$

should have eigenvalues all of whose real parts are negative (section 2.4). This criterion is employed to identify the correct root.

PROBLEM 15. Consider the system described in problem 11, taking the cost function in the form

$$C = \int_0^\infty (x^2 + \tfrac{1}{5}u^2)\, dt.$$

Calculate the optimal control and behaviour from an initial state x_0.

Solution: The matrices are all 1×1 and $A = -2$, $B = 1$, $P = 2$, $Q = 0$, $R = \tfrac{2}{5}$. The steady state Riccati equation takes the form

$$2 - 4K - \tfrac{5}{2}K^2 = 0.$$

Hence, $K = -2$ or $\tfrac{2}{5}$. The corresponding values of the matrix (2.7.30) are $+3$ and -3, respectively; the negative value must be chosen and, thus, $K = \tfrac{2}{5}$.

Equation (2.7.19) is now found to take the form $\dot{x} + 3x = 0$ and it follows that

$$x = x_0 e^{-3t}.$$

Equation (2.7.26) yields the optimal control in the form

$$u = -x = -x_0 e^{-3t}. \qquad \bullet$$

2.8 Escape from a circular orbit

We shall conclude this chapter by studying a problem of optimal escape of a rocket from a circular orbit about a centre of gravitational attraction.

It will be assumed that the motor communicates a known acceleration $f(t)$ to the rocket at time t and that the problem is to choose the direction of thrust in such a manner that the increment in the total energy of the vehicle during a fixed time interval $0 \leqslant t \leqslant T$ is maximized. Since the possibility of escape from a centre of attraction is decided by the value of the total energy, the resulting trajectory will constitute an optimal escape manoeuvre.

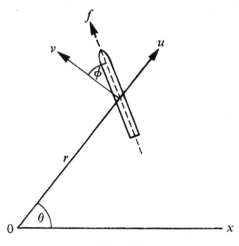

Fig. 2.3

We shall assume that the motion takes place in the plane of the initial circular orbit and the position of the rocket at time t will be specified by polar coordinates (r, θ), with pole at the centre of attraction O (Fig. 2.3). Then, choosing units of length and time such that the radius of the circular orbit and the velocity in this orbit are both unity, $1/r^2$ is the gravitational acceleration at distance r from O and the equations of motion of the vehicle can be written

$$\ddot{r} - r\dot{\theta}^2 = f \sin \phi - \frac{1}{r^2} \tag{2.8.1}$$

$$r\ddot{\theta} + 2\dot{r}\dot{\theta} = f \cos \phi, \tag{2.8.2}$$

where ϕ is the angle of inclination of the thrust to the local horizon. Introducing the radial and transverse components of velocity (u, v), these equations can be reduced to canonical form

$$\dot{r} = u, \qquad \dot{\theta} = v/r, \tag{2.8.3}$$

$$\dot{u} = \frac{v^2}{r} - \frac{1}{r^2} + f \sin \phi, \tag{2.8.4}$$

$$\dot{v} = -\frac{uv}{r} + f \cos \phi. \tag{2.8.5}$$

No use will be made of the second of these equations and this will accordingly, henceforward, be ignored. Hence, the components of

CONTROL SYSTEMS

the state vector x will be taken to be (r, u, v) and the control vector will possess a single component ϕ.

Initially, $t = 0$, and the rocket is moving in a circular orbit $r = 1$ with velocity components $u = 0$, $v = 1$. Thus,

$$x^0 = (1, 0, 1). \qquad (2.8.6)$$

At any instant, the sum of the kinetic and potential energies of the rocket per unit mass is given by

$$E = \tfrac{1}{2}(u^2 + v^2) - \frac{1}{r}. \qquad (2.8.7)$$

If the rocket is to escape, E must be raised to a positive value. Hence, we take the performance index to be

$$C = \tfrac{1}{2}(u_1^2 + v_1^2) - \frac{1}{r_1}, \qquad (2.8.8)$$

where $x^1 = (r_1, u_1, v_1)$ denotes the final state at $t = T$, and choose $\phi(t)$ to maximize C.

The Hamiltonian for the problem is

$$H = \lambda_r u + \lambda_u \left(\frac{v^2}{r} - \frac{1}{r^2} + f \sin \phi\right) + \lambda_v \left(-\frac{uv}{r} + f \cos \phi\right) \qquad (2.8.9)$$

and Hamilton's equations take the form

$$\dot\lambda_r = -\frac{\partial H}{\partial r} = \lambda_u \left(\frac{v^2}{r^2} - \frac{2}{r^3}\right) - \lambda_v \frac{uv}{r^2}, \qquad (2.8.10)$$

$$\dot\lambda_u = -\frac{\partial H}{\partial u} = -\lambda_r + \lambda_v \frac{v}{r}, \qquad (2.8.11)$$

$$\dot\lambda_v = -\frac{\partial H}{\partial v} = -\lambda_u \frac{2v}{r} + \lambda_v \frac{u}{r}. \qquad (2.8.12)$$

The optimal control law is determined by

$$\frac{\partial H}{\partial \phi} = f(\lambda_u \cos \phi - \lambda_v \sin \phi) = 0, \qquad (2.8.13)$$

i.e.

$$\tan \phi = \lambda_u / \lambda_v. \qquad (2.8.14)$$

The end conditions (2.6.14), with G determined by equation (2.8.8), yield

$$\lambda_r = 1/r_1^2, \quad \lambda_u = u_1, \quad \lambda_v = v_1, \qquad (2.8.15)$$

at $t = T$. It follows from these equations and equation (2.8.14) that the thrust and velocity vectors are aligned at the final instant.

Equations (2.8.3)–(2.8.5), (2.8.10)–(2.8.12), (2.8.14) together with the end conditions (2.8.6), (2.8.15), now specify the optimal trajectory of escape. An exact solution in terms of known functions cannot be found and either an approximate solution must be sought or numerical methods resorted to. Numerical integration is difficult since starting values of the six dependent variables are not all available at either of the terminals $t = 0$, $t = T$; it is therefore necessary to take trial values of λ_r, λ_u, λ_v at $t = 0$ and later to adjust these until the end conditions (2.8.15) are satisfied. This process of repeated integration is laborious and the services of a computer are clearly essential if it is to be completed successfully. Most optimization problems of practical interest reduce to mixed end-point integration situations of this type and therein lies their mathematical difficulty; even aided by a large computer, systems of order higher than five (i.e. state vectors with more than five components) are usually intractable.

Assuming that the rocket thrust is developed by a motor which operates by burning conventional propellents, the acceleration f will be large by comparison with the gravitational acceleration $1/r^2$ and an approximate solution can be developed thus: A new independent variable τ is defined by the equation

$$\tau = \int_0^t f(s)\, ds. \qquad (2.8.16)$$

Hence, $d\tau/dt = f$ and equations (2.8.3)–(2.8.5), (2.8.10)–(2.8.12) can be written

$$\left. \begin{aligned} r' &= \frac{u}{f}, \qquad u' = \frac{1}{f}\!\left(\frac{v^2}{r} - \frac{1}{r^2}\right) = \sin \phi, \\ v' &= -\frac{uv}{fr} + \cos \phi, \\ \lambda_r' &= \frac{1}{f}\!\left[\lambda_u\!\left(\frac{v^2}{r^2} - \frac{2}{r^3}\right) - \lambda_v \frac{uv}{r^2}\right], \\ \lambda_u' &= \frac{1}{f}\!\left(-\lambda_r + \lambda_v \frac{v}{r}\right), \\ \lambda_v' &= \frac{1}{f}\!\left(-\lambda_u \frac{2v}{r} + \lambda_v \frac{u}{r}\right), \end{aligned} \right\} \qquad (2.8.17)$$

where primes denote differentiations with respect to τ. We now replace $1/f$ everywhere by ϵ/f and develop a solution in ascending powers of ϵ, putting $\epsilon = 1$ at the final stage.

Thus, putting

$$r = r(0) + \epsilon r(1) + \epsilon^2 r(2) + \cdots, \quad (2.8.18)$$

etc. into equations (2.8.17) (with $1/f$ replaced by ϵ/f) and equating coefficients of like powers of ϵ from the two sides of each equation, we obtain the following set of zero order equations

$$\left.\begin{array}{l} r'(0) = 0, \quad u'(0) = \sin \phi(0), \quad v'(0) = \cos \phi(0), \\ \lambda_r'(0) = \lambda_u'(0) = \lambda_v'(0) = 0, \end{array}\right\} \quad (2.8.19)$$

the following set of first order equations

$$\left.\begin{array}{l} r'(1) = \dfrac{u(0)}{f}, \\[4pt] u'(1) = \dfrac{1}{f}\left(\dfrac{v^2(0)}{r(0)} - \dfrac{1}{r^2(0)}\right) + \phi(1) \cos \phi(0), \\[4pt] v'(1) = -\dfrac{u(0)v(0)}{fr(0)} - \phi(1) \sin \phi(0), \\[4pt] \lambda_r'(1) = \dfrac{1}{f}\left[\lambda_u(0)\left(\dfrac{v^2(0)}{r^2(0)} - \dfrac{2}{r^3(0)}\right) - \lambda_v(0)\dfrac{u(0)v(0)}{r^2(0)}\right], \\[4pt] \lambda_u'(1) = \dfrac{1}{f}\left(-\lambda_r(0) + \lambda_v(0)\dfrac{v(0)}{r(0)}\right), \\[4pt] \lambda_v'(1) = \dfrac{1}{f}\left(-\lambda_u(0)\dfrac{2v(0)}{r(0)} + \lambda_v(0)\dfrac{u(0)}{r(0)}\right), \end{array}\right\} \quad (2.8.20)$$

and so on. Similarly, the optimal control equation (2.8.14) separates into a zero order equation

$$\tan \phi(0) = \lambda_u(0)/\lambda_v(0), \quad (2.8.21)$$

and a first order equation

$$\phi(1) \sec^2 \phi(0) = \dfrac{\lambda_u(1)}{\lambda_v(0)} - \dfrac{\lambda_u(0)\lambda_v(1)}{\lambda_v^2(0)}. \quad (2.8.22)$$

The initial conditions (2.8.6) separate into zero order conditions

$$r(0) = 1, \quad u(0) = 0, \quad v(0) = 1, \quad (2.8.23)$$

and first order conditions
$$r(1) = u(1) = v(1) = 0, \qquad (2.8.24)$$
all applicable at $\tau = 0$. The end conditions (2.8.15) require that, at
$$\tau = \tau_1 = \int_0^T f(s)\,ds, \qquad (2.8.25)$$
$$\lambda_r(0) = 1/r^2(0), \qquad \lambda_u(0) = u(0), \qquad \lambda_v(0) = v(0), \quad (2.8.26)$$
and
$$\lambda_r(1) = -\frac{2r(1)}{r^3(0)}, \qquad \lambda_u(1) = u(1), \qquad \lambda_v(1) = v(1). \quad (2.8.27)$$

Solving the zero order equations (2.8.19), (2.8.21) under the end conditions (2.8.23), (2.8.26), it will be found that
$$\left.\begin{array}{l}\phi(0) = 0, \quad r(0) = 1, \quad u(0) = 0, \quad v(0) = \tau + 1 \\ \lambda_r(0) = 1, \quad \lambda_u(0) = 0, \quad \lambda_v(0) = \tau_1 + 1.\end{array}\right\} \quad (2.8.28)$$

This zero order solution corresponds to the case when f is infinite and, hence, T must be zero if the energy increment is to be finite. In these circumstances, the motor thrust is impulsive and, since $\phi(0) = 0$, must be applied in the direction of the vehicle's motion. This result is precisely the one our intuition leads us to expect.

The results (2.8.28) can now be substituted into the first order equations (2.8.20), (2.8.22) and the unknown first order functions $r(1)$, etc. solved for under the end conditions (2.8.24), (2.8.27). To simplify matters, only the solution for the case $f = $ constant will be given (then $\tau_1 = fT$), viz.

$$r(1) = v(1) = 0,$$
$$u(1) = \frac{1}{2f}\bigg[(\tau + 1)^3 - \frac{1}{\tau_1 + 1}(\tau + 1)^2$$
$$- \frac{\tau_1^2 + 3\tau_1 + 1}{\tau_1 + 1}(\tau + 1) + \tau_1 + 1\bigg], \quad (2.8.29)$$
$$\lambda_r(1) = \lambda_v(1) = 0,$$
$$\lambda_u(1) = \frac{1}{2f}\bigg[(\tau_1 + 1)(\tau + 1)^2 - 2(\tau + 1) - \tau_1^2 - \tau_1 + 1\bigg],$$
$$\phi(1) = \lambda_u(1)/(\tau_1 + 1).$$

This process can now be continued to second and higher orders in ϵ, but the formulae obtained rapidly increase in complexity. However, for the values of f normally met with in practice (between 5 and

10), little improvement results when allowance is made for these higher order terms and the solution which has already been found proves to be adequate for most purposes. To this order, the final energy C defined by equation (2.8.8) will be found to be given by

$$C_{\text{opt}} = \tfrac{1}{2}(\tau_1 + 1)^2 - 1 + \frac{1}{8f^2}\tau_1^4(\tau_1 + 2)^2. \qquad (2.8.30)$$

It is of interest to calculate the "throw-off angle" δ, i.e. the angle made by the thrust vector with the velocity vector, for the optimal manoeuvre. If ψ is the angle of inclination of the velocity vector to the local horizon, then

$$\tan \psi = \frac{u}{v}$$

$$= \frac{1}{2f}\left[(\tau + 1)^2 - \frac{\tau + 1}{\tau_1 + 1} + \frac{\tau_1 + 1}{\tau + 1} - \frac{\tau_1^2 + 3\tau_1 + 1}{\tau_1 + 1}\right], \qquad (2.8.31)$$

to the first order in $1/f$. Since ψ is small, we now calculate that

$$\delta = \psi - \phi = \frac{1}{2f} \cdot \frac{(\tau_1 - \tau)^2}{(\tau_1 + 1)(\tau + 1)}. \qquad (2.8.32)$$

It will be observed that δ is never negative, implying that the thrust must always be thrown off from the direction of motion in the sense towards the centre of attraction.

Exercises 2

1. A system is governed by the second order equation

$$(t^2 - 1)\ddot{x} - 2t\dot{x} + 2x = u.$$

Derive equations in canonical form and the associated state transition matrix. Hence show that the system's response to the control $u = (t^2 - 1)^2$ from an initial state $x = \dot{x} = 0$ at $t = 2$ is given by

$$x = \tfrac{1}{6}(t^4 - 15t^2 + 28t - 12).$$

2. If

$$A = \begin{bmatrix} 3 & 2 \\ 1 & 4 \end{bmatrix},$$

show that

$$\log A = \tfrac{1}{3}(I \log \tfrac{32}{25} + A \log \tfrac{5}{2}).$$

3. If the control equations for a certain system are
$$\dot{x}_1 = x_1 - 3x_2 + u,$$
$$\dot{x}_2 = 2x_1 - 4x_2 - e^{-t}u,$$
calculate the state transition matrix and, hence, calculate the response $x(t)$ to the control $u(t) = e^{-t}$, given that $x_1 = x_2 = 1$ at $t = 0$.
(Ans. $x_1 = (3t + 1)e^{-t} - 3te^{-2t}$; $x_2 = 2te^{-t} - (3t - 1)e^{-2t}$.)

4. A vehicle commences to move from rest at the origin in the xy-plane at $t = 0$. Its acceleration has unit magnitude throughout its motion. If θ is the angle made by the acceleration's direction with the x-axis, (x, y) are the coordinates of the vehicle and (u, v) are its velocity components, obtain θ as a function of t if the quantity $C = y_1 + u_1$ is to be maximized, where $t_1 = 1$. Deduce that the vehicle's path is given by the equations

$$x = \sqrt{2} - \sqrt{[1 + (1 - t)^2]} + (1 - t)[\sinh^{-1}(1 - t) - \sinh^{-1} 1],$$
$$y = \sqrt{2}(t - \tfrac{1}{2}) + \tfrac{1}{2}(1 - t)\sqrt{1 + (1 - t)^2}$$
$$+ \tfrac{1}{2}[\sinh^{-1}(1 - t) - \sinh^{-1} 1].$$

5. A certain system is governed by the equation $\dot{x} = -x + u$. Its initial state is $x = 1$ at $t = 0$ and the final state at $t = 1$ is not specified. $u(t)$ is to be chosen so that

$$C = \int_0^1 (x^2 + u^2/3)\, dt$$

is minimized. Calculate the optimal control and show that the minimal cost is $1/(1 + 2 \coth 2)$.
(Ans. $u = -[3 \sinh 2(1 - t)]/(2 \cosh 2 + \sinh 2)$.)

6. The equation of motion of an oscillatory system is $\ddot{x} + x = u$. Its initial state at $t = 0$ is given by $x = 0$, $\dot{x} = 1$. The control $u(t)$ is to be chosen over the interval $0 \leqslant t \leqslant \tfrac{1}{2}\pi$ so that

$$C = x_1^2 + \int_0^{\frac{1}{2}\pi} u^2\, dt$$

is minimized. Show that $u(t) = (-4 \cos t)/(\pi + 4)$ and that $C_{\min} = 4/(\pi + 4)$.

7. A system is governed by the following equations of motion: $\dot{x} = y$, $\dot{y} = u$, u being the control. If the system is disturbed from the equilibrium state $x = y = 0$ at $t = 0$, control should be such as

to minimize
$$\int_0^\infty (x^2 + u^2)\, dt.$$
Show that the control equation which realises this control is $u = -x - \sqrt{2}y$.

8. Obtain the Riccati equation for the problem stated in Ex. 5 and deduce that
$$K = \frac{1 - e^{-4(1-t)}}{3 + e^{-4(1-t)}}.$$
Hence solve the problem.

9. A system has the equation of motion $\ddot{x} + x = u$ and its initial state at $t = 0$ is given by $x = 0$, $\dot{x} = 1$. The cost over the time of operation $0 \leqslant t < \infty$ is given by
$$C = \int_0^\infty (x^2 + u^2/8)\, dt.$$
Minimize C by showing that the solution of the steady state Riccati equation is
$$K = \tfrac{1}{2}\begin{bmatrix} 3 & 1 \\ 1 & 1 \end{bmatrix}.$$
Deduce that $u = -2e^{-t}\cos\sqrt{2}t$ and that the minimal value of C is $\tfrac{1}{4}$.

10. A Duffing oscillator is governed by the non-linear equation $\ddot{x} + x + \epsilon x^3 = u$. At $t = 0$, $x = 0$, $\dot{x} = 1$. $u(t)$ is to be chosen so that
$$C = \int_0^\infty (x^2 + u^2/8)\, dt$$
is minimized. Assuming that ϵ is sufficiently small so that the state and control variables and the Lagrange multipliers can all be expanded in series of powers of ϵ, show that to the first order in ϵ
$$u = -2e^{-t}\cos\sqrt{2}t + \epsilon[\tfrac{1}{432}e^{-t}(-13\cos\sqrt{2}t + 10\sqrt{2}\sin\sqrt{2}t)$$
$$+ \tfrac{1}{32}e^{-3t}(2\cos\sqrt{2}t - \sqrt{2}\sin\sqrt{2}t)$$
$$+ \tfrac{1}{864}e^{-3t}(8\cos 3\sqrt{2}t + 17\sqrt{2}\sin 3\sqrt{2}t)] + O(\epsilon^2),$$
$$x = \frac{1}{\sqrt{2}}e^{-t}\sin\sqrt{2}t + \epsilon[\tfrac{1}{1728}e^{-t}(20\cos\sqrt{2}t + 13\sqrt{2}\sin\sqrt{2}t)$$
$$- \tfrac{1}{64}e^{-3t}(\cos\sqrt{2}t + \sqrt{2}\sin\sqrt{2}t)$$
$$+ \tfrac{1}{1728}e^{-3t}(7\cos 3\sqrt{2}t - 2\sqrt{2}\sin 3\sqrt{2}t] + O(\epsilon^2).$$

3 | ADDITIONAL CONSTRAINTS

3.1 Constraints on the final state

In the previous chapter, it was assumed that the final state x^1 of the control system, whose behaviour was to be optimized, could be chosen quite freely. We shall now study problems in which this final state is either completely specified, or is subject to some degree of constraint. At the same time, we shall relax the constraint on t^1 and suppose that the final instant can be chosen to have any value ($>t^0$) we please.

Thus, suppose the components x_i^1 of the state vector at the final instant t^1 are required to satisfy Q constraints in the form of equations

$$F_m(x_1^1, x_2^1, \ldots, x_M^1, t^1) = 0, \qquad (3.1.1)$$

where $m = 1, 2, \ldots, Q$. It will be assumed that these constraints are independent and that $Q \leqslant M + 1$. If $Q = M + 1$, equations (3.1.1) will possess a finite number of solutions for x^1, t^1 only, and it will then be a requirement that the control $u(t)$ shall always be chosen in such a manner that the system is driven into one of these final states at a predetermined time t^1. If $Q < M + 1$, neither the final state nor the final time are necessarily predetermined. It will invariably be supposed that controls $u(t)$ exist which are capable of satisfying the constraints (3.1.1); such controls will be said to be *admissible*. Our problem is to identify within the set of admissible controls, that which leads to the cost function C taking a minimum value. We shall define C by the equation

$$C = G(x^1, t^1) + \int_{t^0}^{t^1} g(x, u, t) \, dt \qquad (3.1.2)$$

and suppose that the initial state x^0 of the system at time t^0 is fixed.

If the system has the property that it can always be driven from a given initial state x^0 at $t = t^0$ to any arbitrarily chosen final state x^1 for some value of t^1, it is said to be *completely controllable* at the instant t^0. If the system is autonomous and it is completely controllable for one value of t^0, it is clearly completely controllable for all values of t^0. Another concept which proves to be useful in discussions of the behaviour of control systems is the *set of attainability* Ω. For a given initial state x^0 at $t = t^0$, this is the set of all points x^1 in the state space representing final states into which the system can be driven at a given final instant $t = t^1$. Thus, $\Omega = \Omega(x^0, t^0, t^1)$. In our subsequent argument, we shall assume that the optimal final state which satisfies the constraints (3.1.1) is an interior point of the associated set of attainability; this will imply that the optimal control can be subjected to arbitrary small variations to yield a family of admissible controls within which the optimal control is embedded.

Denoting the optimal control by $u(t)$ and the optimal state behaviour by $x(t)$, the argument of section 2.6 must now be amended to allow for the additional constraints (3.1.1).

Consider a neighbouring non-optimal control determined by the control vector $v(t, \epsilon_1, \epsilon_2, \ldots, \epsilon_{Q+1})$, where

$$v_r = u_r(t) + \epsilon_n \xi_{nr}(t), \tag{3.1.3}$$

$n = 1, 2, \ldots, Q + 1$. In this expression, the functions $\xi_{nr}(t)$ are continuous, but otherwise arbitrary and the quantities ϵ_n are, for the moment, independent parameters. In general, such a control will be inadmissible, since it will not bring the system into a final state which satisfies the constraints (3.1.1). However, we shall later require the parameters ϵ_n to be so related that the control $v(t)$ is admissible. Clearly, $v(t, 0, 0, \ldots, 0) = u(t)$. Let $y(t, \epsilon_1, \epsilon_2, \ldots, \epsilon_{Q+1})$ be the state vector response to v. Then $y(t, 0, 0, \ldots, 0) = x(t)$.

It follows from equation (3.1.3) that

$$\frac{\partial v_r}{\partial \epsilon_n} = \xi_{nr}. \tag{3.1.4}$$

ξ_{nr} is called the *variation* of u_r with respect to the parameter ϵ_n. Similarly (cf. equation (2.6.4)), the variation η_{ni} of x_i with respect to the parameter ϵ_n is defined by the equation

$$\eta_{ni} = \left(\frac{\partial y_i}{\partial \epsilon_n}\right)_{\epsilon_1 = \epsilon_2 = \ldots = \epsilon_{Q+1} = 0}. \tag{3.1.5}$$

We assume that the initial state x^0 of the system at time t^0 is predetermined and is the same for all controls y. Hence

$$y(t^0, \epsilon_1, \epsilon_2, \ldots, \epsilon_{Q+1}) = x^0 \tag{3.1.6}$$

is an identity with respect to the parameters ϵ_n. Partial differentiation with respect to these parameters, followed by the setting of all their values to zero, yields the condition

$$\eta_{ni}^0 = 0. \qquad (3.1.7)$$

The state equation (2.1.2) of the system must be satisfied by $y(t, \epsilon)$ for all (sufficiently small) values of the parameters. Thus,

$$\frac{\partial y}{\partial t} = f(y, v, t). \qquad (3.1.8)$$

Partially differentiating this equation with respect to ϵ_n and putting the parameter values to zero, we get,

$$\dot{\eta}_{ni} = \frac{\partial f_i}{\partial x_j}\eta_{nj} + \frac{\partial f_i}{\partial u_r}\xi_{nr} \qquad (3.1.9)$$

(cf. equation (2.6.9)). These are the equations of variation which, together with the initial conditions (3.1.7), determine the state variations η_{ni} when the control variations ξ_{nr} are given.

Employing equation (3.1.2), the cost associated with the control v can be expressed as a function of the parameters ϵ_n and of the final time t^1, thus:

$$C(\epsilon, t^1) = G(y^1, t^1) + \int_{t^0}^{t^1} g(y, v, t)\, dt. \qquad (3.1.10)$$

However, the parameters ϵ_n and time t^1 are not independent, since they must be chosen to satisfy the constraints (3.1.1). This leads to the Q relationships

$$F_m[y_1(t^1, \epsilon), y_2(t^1, \epsilon), \ldots, y_M(t^1, \epsilon), t^1] = 0 \qquad (3.1.11)$$

connecting the $Q + 2$ quantities $\epsilon_1, \epsilon_2, \ldots, \epsilon_{Q+1}, t^1$. The problem posed is accordingly that of minimizing $C(\epsilon, t^1)$ subject to the Q constraints (3.1.11) upon the variables ϵ, t^1; the solution to this problem is known to be such that $\epsilon = 0$ and t^1 takes its optimal value. This type of problem has been studied in section 1.3 and we shall make use of the necessary conditions (1.3.12) to derive necessary conditions for our present problem.

Introducing Lagrange multipliers ν_m, we first construct the Hamiltonian expression

$$K(\epsilon, \nu, t^1) = C(\epsilon, t^1) + \nu_m F_m(y^1, t^1). \qquad (3.1.12)$$

ADDITIONAL CONSTRAINTS

Then, necessary conditions for the control $u(t)$ to be optimal are that

$$\frac{\partial K}{\partial \epsilon_n} = \frac{\partial K}{\partial t^1} = 0 \qquad (3.1.13)$$

when $\epsilon = (0, 0, \ldots, 0)$ and t^1 takes its optimal value. Partially differentiating the terms of K with respect to ϵ_n and t^1, and subsequently putting $\epsilon = 0$, it will be found that

$$\left(\frac{\partial K}{\partial \epsilon_n}\right)_0 = \frac{\partial G}{\partial x_i^1}\eta_{ni}^1 + \int_{t^0}^{t^1}\left(\frac{\partial g}{\partial x_i}\eta_{ni} + \frac{\partial g}{\partial u_r}\xi_{nr}\right)dt + \nu_m \frac{\partial F_m}{\partial x_i^1}\eta_{ni}^1 \qquad (3.1.14)$$

$$\left(\frac{\partial K}{\partial t^1}\right)_0 = \frac{\partial G}{\partial x_i^1}\dot{x}_i^1 + \frac{\partial G}{\partial t^1} + g(x^1, u^1, t^1) + \nu_m\left(\frac{\partial F_m}{\partial x_i^1}\dot{x}_i^1 + \frac{\partial F_m}{\partial t_1}\right), \qquad (3.1.15)$$

where

$$\dot{x}_i^1 = \left[\frac{\partial}{\partial t}y(t, \epsilon)\right]_{\epsilon=0, t=t^1}. \qquad (3.1.16)$$

[N.B. $y^1 = y(t^1, \epsilon)$ is a function of t^1.]

The expression (3.1.14), is now reduced to a more tractable form by the introduction of multipliers $\lambda_i(t)$ satisfying the adjoint equations

$$\dot{\lambda}_i = -\lambda_j \frac{\partial f_j}{\partial x_i} - \frac{\partial g}{\partial x_i}, \qquad (3.1.17)$$

(cf. section 2.6). End conditions will be imposed upon the multipliers at a later stage. It now follows from equations (3.1.9) and (3.1.17) that

$$\frac{d}{dt}(\lambda_i \eta_{ni}) = \lambda_i \frac{\partial f_i}{\partial u_r}\xi_{nr} - \eta_{ni}\frac{\partial g}{\partial x_i}, \qquad (3.1.18)$$

(cf. equation (2.6.15)). Thus, equation (3.1.14) can be written in the form

$$\left(\frac{\partial K}{\partial \epsilon_n}\right)_0 = \frac{\partial J}{\partial x_i^1}\eta_{ni}^1 + \int_{t^0}^{t^1}\left[\frac{\partial H}{\partial u_r}\xi_{nr} - \frac{d}{dt}(\lambda_i \eta_{ni})\right]dt, \qquad (3.1.19)$$

where

$$H(x, u, \lambda, t) = g + \lambda_i f_i, \qquad (3.1.20)$$

$$J(x^1, \nu, t^1) = G + \nu_m F_m, \qquad (3.1.21)$$

are Hamiltonians. Integrating the second term in the integrand in equation (3.1.19) and using the initial conditions (3.1.7), we can now express the first of the necessary conditions (3.1.13) in the form

$$\left(\frac{\partial J}{\partial x_i^1} - \lambda_i^1\right)\eta_{ni}^1 + \int_{t^0}^{t^1} \frac{\partial H}{\partial u_r} \xi_{nr}\, dt = 0. \quad (3.1.22)$$

At this stage, we choose the end conditions on the multipliers λ_i to be

$$\lambda_i^1 = \frac{\partial J}{\partial x_i^1}, \quad (3.1.23)$$

so that the condition (3.1.22) reduces finally to

$$\int_{t^0}^{t^1} \frac{\partial H}{\partial u_r} \xi_{nr}\, dt = 0. \quad (3.1.24)$$

Since the functions ξ_{nr} are arbitrary, it now follows that

$$\frac{\partial H}{\partial u_r} = 0 \quad (3.1.25)$$

identically along an optimal trajectory.

It remains to calculate the form of the second condition (3.1.13). Since $\dot{x}_i = f_i(x, u, t)$ by the state equations, equation (3.1.15) is equivalent to

$$\left(\frac{\partial K}{\partial t^1}\right)_0 = \frac{\partial J}{\partial x_i^1} f_i^1 + \frac{\partial J}{\partial t^1} + g^1. \quad (3.1.26)$$

Making use of equation (3.1.23), this can be written

$$\left(\frac{\partial K}{\partial t^1}\right)_0 = H^1 + \frac{\partial J}{\partial t^1} \quad (3.1.27)$$

and the second condition (3.1.13) accordingly takes the form

$$H^1 = -\frac{\partial J}{\partial t^1}. \quad (3.1.28)$$

To summarise, along an optimal trajectory it is necessary that

$$\dot{x}_i = \frac{\partial H}{\partial \lambda_i}, \qquad \dot{\lambda}_i = -\frac{\partial H}{\partial x_i}, \qquad \frac{\partial H}{\partial u_r} = 0, \quad (3.1.29)$$

and, at the final instant $t = t^1$, it is necessary that

$$\lambda_i^1 = \frac{\partial J}{\partial x_i^1}, \qquad H^1 = -\frac{\partial J}{\partial t^1}. \quad (3.1.30)$$

In the case when t^1 is predetermined, the condition $H = -\partial J/\partial t^1$ is not applicable. The $(2M + N)$ equations (3.1.29) determine the same number of functions $x_i(t)$, $\lambda_i(t)$, $u_r(t)$ except for $2M$ constants of integration; these integration constants, together with the Q multipliers ν_m and the time t^1 are determined by the M initial conditions on the x_i, the Q end constraints (3.r.1) and the $M + 1$ end conditions (3.1.30).

In the special case when the end constraints (3.1.1) simply specify the final values of the first Q components of the state vector, we have

$$F_m = x_m^1 - X_m, \qquad (3.1.31)$$

where X_m are the given values. Thus $J = G + \nu_m(x_m^1 - X_m)$ and the end conditions (3.1.30) become

$$\lambda_i^1 = \frac{\partial G}{\partial x_i^1} + \nu_i, \qquad H^1 = -\frac{\partial G}{\partial t^1}, \qquad (3.1.32)$$

where ν_i is to be put equal to zero if $i > Q$. The first set of equations with $i = 1, 2, \ldots, Q$ serve merely to fix the multipliers ν_m and, since these occur in none of the remaining equations fixing the optimal trajectory, they may be disregarded. The end conditions at $t = t^1$ therefore reduce to the set

$$\lambda_i^1 = \frac{\partial G}{\partial x_i^1} \ (i = Q + 1, \ldots, M), \qquad H^1 = -\frac{\partial G}{\partial t^1}, \quad (3.1.33)$$

and the Hamiltonian J need not be constructed. Note that only those λ_i associated with components of x which are not predetermined have to satisfy end conditions at $t = t^1$.

PROBLEM 16. A mass hanging vertically by a spring has equation of motion

$$\ddot{x} + x = u,$$

x being its displacement from the equilibrium position and u being an applied force. Initially, $t = 0$, $x = 0$, $\dot{x} = 1$. Calculate $u(t)$ if

$$C = \int_0^{\frac{1}{2}\pi} u^2 \, dt$$

is to be minimized, and x is to vanish at $t = \frac{1}{2}\pi$.

Solution: Replace the equation of motion by the canonical system

$$\dot{x} = y, \qquad \dot{y} = -x + u,$$

and form the Hamiltonian

$$H = u^2 + \lambda_x y + \lambda_y(-x + u).$$

Equations for the optimal control are,

$$\dot\lambda_x = -\frac{\partial H}{\partial x} = \lambda_y,$$

$$\dot\lambda_y = -\frac{\partial H}{\partial y} = -\lambda_x,$$

$$\frac{\partial H}{\partial u} = 2u + \lambda_y = 0.$$

$t^1 = \tfrac{1}{2}\pi$ is fixed, so that the end conditions (3.1.33) reduce to $\lambda_y = 0$ (N.B. y is not predetermined at $t = \tfrac{1}{2}\pi$).

It now follows that $u = -\tfrac{1}{2}\lambda_y = -\tfrac{1}{2}A \cos t$ and the equation of motion for the mass takes the form

$$\ddot x + x = -\tfrac{1}{2}A \cos t.$$

Integrating this equation under the initial conditions $x = 0$, $\dot x = 1$, we obtain

$$x = (1 - \tfrac{1}{4}At)\sin t.$$

A must now be chosen such that $x = 0$ at $t = \tfrac{1}{2}\pi$; thus, $A = 8/\pi$ and

$$x = (1 - 2t/\pi)\sin t$$

yields the optimal behaviour.

The optimal control is

$$u = -\tfrac{1}{2}A \cos t = -\frac{4}{\pi}\cos t$$

and this gives an optimal cost of $C_{\text{opt}} = 4/\pi$. ●

PROBLEM 17. In the 4-terminal network described in Problem 11, the output quantity x is to be increased from an initial value x_0 to a final value x_1 in such a way that

$$C = \int_0^T u^2 \, dt$$

is minimized. Assuming T is open to choice, determine the optimal control.

Solution: The state equation has been found as equation (2.6.23); it follows that the Hamiltonian is given by
$$H = u^2 + \lambda(-2x + u),$$
and Hamilton's equations are accordingly
$$\dot{x} = \frac{\partial H}{\partial \lambda} = -2x + u,$$
$$\dot{\lambda} = -\frac{\partial H}{\partial x} = 2\lambda.$$
The optimal control law is determined by the equation
$$\frac{\partial H}{\partial u} = 2u + \lambda = 0.$$
Eliminating u and λ between these equations, we find that
$$\ddot{x} - 4x = 0,$$
an equation which has the general solution
$$x = Ae^{2t} + Be^{-2t}.$$
λ and u can now be derived:
$$\lambda = -8Ae^{2t}, \quad u = 4Ae^{2t}.$$

Substituting for x, λ, u into the Hamiltonian, we calculate that $H = 16AB$. The fact that H is constant is not fortuitous, but will be explained in the next section. According to the second condition (3.1.33), H must vanish at $t = T$; we conclude that $B = 0$ (A cannot vanish, since this would imply that x is decreasing). The end conditions on x now lead to the equations
$$x_0 = A, \quad x_1 = Ae^{2T},$$
from which it follows that
$$T = \tfrac{1}{2} \log(x_1/x_0).$$
The optimal behaviour and control are now given by the equations
$$x = x_0 e^{2t}, \quad u = 4x_0 e^{2t}. \qquad \bullet$$

PROBLEM 18. A rocket is moving in an orbit about the earth with its motor inactive. At a prescribed instant, propellent is fed to the motor and thereafter the thrust programme is determined. It is required to transfer the rocket into some other specified orbit in the minimum time. Show how to calculate the optimal thrust direction programme.

Solution: It will be assumed that the initial and final orbits lie in the same plane through the earth's centre O and, hence, that the transfer trajectory also lies in this plane. Then, defining polar coordinates (r, θ) as in section 2.8, the equations of motion of the rocket can be written in the canonical forms (2.8.3)–(2.8.5).

The initial values of the state variables (r, θ, u, v) are known and their final values must be such that the rocket is injected into the required orbit. Suppose

$$r_1 = P(\theta_1) \tag{3.1.34}$$

is the polar equation of this orbit and that the velocity components of a body moving in the orbit are given by

$$u_1 = Q(\theta_1), \qquad v_1 = R(\theta_1). \tag{3.1.35}$$

The forms of the functions P, Q, R for the case of an inverse square law elliptical orbit are well-known, but will not be quoted here. Equations (3.1.34), (3.1.35) provide constraints on the final state.

The cost function to be minimized is given by

$$C = \int_{t_0}^{t_1} dt \tag{3.1.36}$$

and the Hamiltonian for the problem is accordingly

$$H = 1 + \lambda_r u + \lambda_\theta \frac{v}{r} + \lambda_u \left(\frac{v^2}{r} - \frac{1}{r^2} + f \sin \phi \right)$$
$$+ \lambda_v \left(-\frac{uv}{r} + f \cos \phi \right). \tag{3.1.37}$$

Hamilton's equations can now be written down thus:

$$\left.\begin{aligned}
\dot\lambda_r &= \lambda_\theta \frac{v}{r^2} + \lambda_u \left(\frac{v^2}{r^2} - \frac{2}{r^3} \right) - \lambda_v \frac{uv}{r^2}, \\
\dot\lambda_\theta &= 0, \qquad \dot\lambda_u = -\lambda_r + \lambda_v \frac{u}{r}, \\
\dot\lambda_v &= -\lambda_\theta \frac{1}{r} - \lambda_u \frac{2v}{r} + \lambda_v \frac{u}{r}.
\end{aligned}\right\} \tag{3.1.38}$$

After these equations have been integrated, the optimal thrust direction programme follows from the equation

$$\frac{\partial H}{\partial \phi} = f(\lambda_u \cos \phi - \lambda_v \sin \phi) = 0;$$

i.e.

$$\tan \phi = \lambda_u/\lambda_v. \qquad (3.1.39)$$

The Hamiltonian for the end constraints is

$$J = \nu_r[r_1 - P(\theta_1)] + \nu_u[u_1 - Q(\theta_1)] + \nu_v[v_1 - R(\theta_1)] \quad (3.1.40)$$

and the end conditions to be satisfied by the multipliers at $t = t_1$ (equations (3.1.30)) take the form

$$\left. \begin{aligned} \lambda_{r1} &= \frac{\partial J}{\partial r_1} = \nu_r, \\ \lambda_{\theta 1} &= \frac{\partial J}{\partial \theta_1} = -\nu_r P'(\theta_1) - \nu_u Q'(\theta_1) - \nu_v R'(\theta_1), \\ \lambda_{u1} &= \frac{\partial J}{\partial u_1} = \nu_u, \qquad \lambda_{v1} = \frac{\partial J}{\partial v_1} = \nu_v. \end{aligned} \right\} \quad (3.1.41)$$

Also, since t_1 is variable, there is a further end condition

$$H_1 = \frac{\partial J}{\partial t_1} = 0. \qquad (3.1.42)$$

Elimination of the ν-multipliers between equations (3.1.41), reduces these to the single end condition

$$\lambda_{\theta 1} + \lambda_{r1} P'(\theta_1) + \lambda_{u1} Q'(\theta_1) + \lambda_{v1} R'(\theta_1) = 0. \qquad (3.1.43)$$

The four end conditions (3.1.34), (3.1.35), (3.1.43), serve to determine the four integration constants which arise as a result of integrating equations (3.1.38); the terminal $t = t_1$ of integration is fixed by equation (3.1.42). The final calculation must, of course, be done numerically, employing a computer. Initial trial values are taken for the λ-multipliers and the equations of motion and equations (3.1.38), (3.1.39) integrated to a value $t = t_1$ at which H vanishes. Satisfaction of the four end conditions can now be tested and, assuming failure, the trial values of the multipliers must be adjusted, and the integration repeated, until agreement is obtained.

ANALYTICAL METHODS OF OPTIMIZATION

Since H is not explicitly dependent upon t, as will be proved in the next section, the equation (3.1.42) is valid over the whole optimal trajectory. The existence of this first integral results in some simplification of the numerical procedure but, nevertheless, it clearly remains complex. ●

3.2 An integral of Hamilton's equations

In certain circumstances, a first integral of Hamilton's equations can be found, thus: Differentiating the Hamiltonian (3.1.20) totally with respect to t, it follows that

$$\frac{dH}{dt} = \frac{\partial H}{\partial x_i}\dot{x}_i + \frac{\partial H}{\partial u_j}\dot{u}_j + \frac{\partial H}{\partial \lambda_i}\dot{\lambda}_i + \frac{\partial H}{\partial t} = \frac{\partial H}{\partial t}, \qquad (3.2.1)$$

having used equations (2.6.20)–(2.6.22), valid on the optimal trajectory. If, now, as is frequently the case, H is not explicitly dependent upon t, then $\partial H/\partial t$ vanishes and equation (3.2.1) has an integral $H =$ constant. The value of this constant follows from the end condition (3.1.28) (or (3.1.33)) (assuming this is applicable, i.e. t^1 is variable) and the first integral we are seeking accordingly takes the form

$$H = -\frac{\partial J}{\partial t^1}\left(\text{or } -\frac{\partial G}{\partial t^1}\right). \qquad (3.2.2)$$

PROBLEM 19. Two points on the earth's surface are connected by a smooth tunnel running through its interior. The tunnel is evacuated and a body is allowed to fall through it under gravity. Calculate the shape of the tunnel if the time of transit between the points is a minimum.

Solution: It will be assumed that the tunnel lies in the plane of the great circle passing through the terminals and that the acceleration due to gravity at a distance r from the earth's centre O is $\omega^2 r$ towards O (i.e. the earth is assumed to be a homogeneous sphere).

When the body is at a point P (Fig. 3.1) in the tunnel, let v be its speed and let ϕ be the angle between the axis of the tunnel and the radius $OP = r$ from the earth's centre. θ is the angle made by OP with a fixed reference direction OX. Since the force R exerted by the smooth tunnel wall upon the body is perpendicular to the direction of motion, the body's equation of motion resolved in the direction

ADDITIONAL CONSTRAINTS

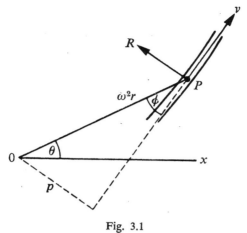

Fig. 3.1

of v is

$$\dot{v} = -\omega^2 r \cos \phi. \quad (3.2.3)$$

The radial and transverse velocity components are given by the equations

$$\dot{r} = v \cos \phi, \qquad r\dot{\theta} = v \sin \phi. \quad (3.2.4)$$

The state vector is now defined to have components (v, r, θ) and the control vector (determining the shape of the tunnel) to have a single component ϕ; thus, equations (3.2.3), (3.2.4) are the state equations in the canonical form.

The state vector is known at both terminals (v vanishes at both these points) and the problem is to choose $\phi(t)$ in such a way that the time of transit is minimized. Thus, we take the cost to be given by

$$C = \int_{t^0}^{t^1} dt; \quad (3.2.5)$$

in the nomenclature of the general theory, this means that $g \equiv 1$, $G \equiv 0$.

The Hamiltonian can now be constructed, thus:

$$H = 1 - \lambda_v \omega^2 r \cos \phi + \lambda_r v \cos \phi + \lambda_\theta \frac{v}{r} \sin \phi. \quad (3.2.6)$$

This leads to a set of Hamilton equations

$$\dot{\lambda}_v = -\frac{\partial H}{\partial v} = -\lambda_r \cos \phi - \frac{\lambda_\theta}{r} \sin \phi,$$

$$\dot{\lambda}_r = -\frac{\partial H}{\partial r} = \lambda_v \omega^2 \cos \phi + \lambda_\theta \frac{v}{r^2} \sin \phi,$$

$$\dot{\lambda}_\theta = -\frac{\partial H}{\partial \theta} = 0.$$

Thus $\lambda_\theta = A$ (a constant). Since all components of the state vector are known at the terminals, there are no end conditions on the multipliers.

The optimal control equation is

$$\frac{\partial H}{\partial \phi} = \lambda_v \omega^2 r \sin \phi - \lambda_r v \sin \phi + \lambda_\theta \frac{v}{r} \cos \phi = 0.$$

This implies that

$$\lambda_r v - \lambda_v \omega^2 r = \mu \cos \phi, \qquad \lambda_\theta \frac{v}{r} = \mu \sin \phi, \qquad (3.2.7)$$

where μ is a proportionality factor.

The first integral (3.2.2) exists and takes the form $H = 0$. Substituting from equations (3.2.7) into this integral, it reduces to $1 + \mu = 0$. Thus, equations (3.2.7) give

$$\lambda_r v - \lambda_v \omega^2 r = -\cos \phi, \qquad \frac{Av}{r} = -\sin \phi. \qquad (3.2.8)$$

The equation of motion (3.2.3) is equivalent to the equation

$$v\dot{v} = -\omega^2 rv \cos \phi = -\omega^2 r\dot{r}. \qquad (3.2.9)$$

This integrates immediately to yield

$$v^2 = \omega^2(R^2 - r^2), \qquad (3.2.10)$$

since $v = 0$ when $r = R$ (earth's radius) at the starting point. Coupling this equation with the second of the optimal control equations (3.2.8), it follows that

$$r^2 \sin^2 \phi = A^2 \omega^2 (R^2 - r^2). \qquad (3.2.11)$$

ADDITIONAL CONSTRAINTS

Now let p ($= r \sin \phi$) be the perpendicular from O on to the line of motion at P. Then, the last equation can be written in the form

$$r^2 = R^2 - \frac{p^2}{A^2\omega^2}. \qquad (3.2.12)$$

This is the pr-equation of the body's optimal trajectory and is well-known (see e.g. *An Elementary Treatise on Cubic and Quartic Curves* by A. B. Basset, Cambridge, 1901, p. 208) to correspond to a hypocycloid traced out by a point on the circumference of a disc which rolls on the inside of a circle of radius R and centre O (i.e. the earth's surface). The radius of the rolling disc is fixed by the constant $A^2\omega^2$ and has to be chosen so that the curve joins the given terminals. If a is this radius, it is known that

$$A^2\omega^2 = \frac{(R - 2a)^2}{4a(R - a)}. \qquad (3.2.13)$$

Let α be the angle subtended at O by the minor great circle arc joining the terminals. Then the circumference of the rolling disc must have length $R\alpha$ and its radius must therefore be $a = R\alpha/2\pi$. Substituting for a into equation (3.2.13), we find that

$$A^2\omega^2 = \frac{(\pi - \alpha)^2}{\alpha(2\pi - \alpha)}. \qquad (3.2.14)$$

To calculate the time of transit, we use the first of equations (3.2.4). Equations (3.2.10), (3.2.11), show that this is equivalent to the equation

$$\frac{dr}{dt} = \frac{\omega}{r}\left[(R^2 - r^2)\{(1 + A^2\omega^2)r^2 - A^2\omega^2 R^2\}\right]^{\frac{1}{2}}. \qquad (3.2.15)$$

This equation indicates that r first decreases from its initial value R to a minimum value $A\omega R(1 + A^2\omega^2)^{-\frac{1}{2}} = R_0$ (when $\dot{r} = 0$); it then increases, in a symmetrical manner, to its final value R again. Thus, by integration of equation (3.2.15), we conclude that the transit time is

$$\frac{2}{\omega\sqrt{(1 + A^2\omega^2)}} \int_{R_0}^{R} \frac{r\,dr}{\sqrt{[(R^2 - r^2)(r^2 - R_0^2)]}}$$
$$= \frac{\pi}{\omega\sqrt{(1 + A^2\omega^2)}} = \frac{1}{\omega}\sqrt{[\alpha(2\pi - \alpha)]}, \qquad (3.2.16)$$

having used equation (3.2.14). It will be seen that the maximum transit time occurs when $\alpha = \pi$; the optimal trajectory is then a diameter of the earth and the transit time is π/ω. ●

PROBLEM 20. Use *Fermat's Principle of Least Time* to derive differential equations for the path of a light ray through an optical medium having variable refractive index μ.

Solution: Fermat's principle states that the light ray joining two fixed points A and B in an optical medium is such that the time it takes a pulse of light to traverse it is less than for all neighbouring paths joining the points.

Only the problem in a plane will be studied. Let (x, y) be cartesian coordinates relative to a rectangular frame and let $\mu(x, y)$ be the refractive index at any point (x, y) in the medium. Consider a light pulse which moves along some path joining A and B, such that its speed at any point on the path is equal to the speed of propagation of light at the point, viz. c/μ, where c is the speed of light in vacuo. Then, if θ is the angle made by the tangent to the path with the x-axis, the equations of motion of the pulse are

$$\dot{x} = \frac{c}{\mu} \cos \theta, \qquad \dot{y} = \frac{c}{\mu} \sin \theta. \tag{3.2.17}$$

We now treat (x, y) as state variables and θ as the control variable and wish to minimize the path time

$$C = \int_{t_0}^{t_1} dt. \tag{3.2.18}$$

Thus, the Hamiltonian for the problem is given by

$$H = 1 + \frac{c}{\mu}(\lambda_x \cos \theta + \lambda_y \sin \theta) \tag{3.2.19}$$

and the optimal control is determined by the equation

$$\frac{\partial H}{\partial \theta} = \frac{c}{\mu}(-\lambda_x \sin \theta + \lambda_y \cos \theta) = 0,$$

or

$$\tan \theta = \lambda_y/\lambda_x. \tag{3.2.20}$$

ADDITIONAL CONSTRAINTS

Hamilton's equations are

$$\dot{\lambda}_x = -\frac{\partial H}{\partial x} = \frac{c}{\mu^2}\frac{\partial \mu}{\partial x}(\lambda_x \cos\theta + \lambda_y \sin\theta),$$

$$\dot{\lambda}_y = -\frac{\partial H}{\partial y} = \frac{c}{\mu^2}\frac{\partial \mu}{\partial y}(\lambda_x \cos\theta + \lambda_y \sin\theta).$$

Since the values of x and y are known at the terminal B, there are no end conditions upon λ_x and λ_y.

H is not explicitly dependent upon t and t_1 is variable. It follows that $H = 0$ is a first integral, and Hamilton's equations accordingly reduce to the form

$$\dot{\lambda}_x = -\frac{1}{\mu}\frac{\partial \mu}{\partial x}, \qquad \dot{\lambda}_y = -\frac{1}{\mu}\frac{\partial \mu}{\partial y}. \tag{3.2.21}$$

Also, the integral $H = 0$ and equations (3.2.20) are equivalent to the equations

$$\lambda_x \cos\theta + \lambda_y \sin\theta = -\mu/c,$$
$$\lambda_x \sin\theta - \lambda_y \cos\theta = 0.$$

These imply that

$$\left.\begin{aligned}\lambda_x &= -\frac{\mu}{c}\cos\theta = -\frac{\mu}{c}\frac{dx}{ds}, \\ \lambda_y &= -\frac{\mu}{c}\sin\theta = -\frac{\mu}{c}\frac{dy}{ds},\end{aligned}\right\} \tag{3.2.22}$$

where s is arc length measured along the path from A. Thus, equations (3.2.21) can be expressed in the form

$$\frac{d}{dt}\left(\mu\frac{dx}{ds}\right) = \frac{c}{\mu}\frac{\partial \mu}{\partial x}, \qquad \frac{d}{dt}\left(\mu\frac{dy}{ds}\right) = \frac{c}{\mu}\frac{\partial \mu}{\partial y}. \tag{3.2.23}$$

But, equations (3.2.17) show that

$$\dot{s} = \sqrt{(\dot{x}^2 + \dot{y}^2)} = c/\mu$$

and equations (3.2.23) can therefore be written in their final form

$$\frac{d}{ds}\left(\mu\frac{dx}{ds}\right) = \frac{\partial \mu}{\partial x}, \qquad \frac{d}{ds}\left(\mu\frac{dy}{ds}\right) = \frac{\partial \mu}{\partial y}. \tag{3.2.24}$$

Integration of the last pair of equations under the end conditions imposed by A and B leads to parametric equations for the light ray in the form $x = x(s)$, $y = y(s)$. ●

3.3 Calculus of variations problem

The fundamental problem which initiated the study of the calculus of variations was that of minimizing an integral of the form

$$\int_{x_0}^{x_1} F[x, y(x), y'(x)]\, dx$$

with respect to the function $y(x)$. To be admissible, $y(x)$ was required to possess a continuous first derivative and the end points $A(x_0, y(x_0))$, $B(x_1, y(x_1))$ were required to lie on two given curves Γ_0, Γ_1 in the xy-plane (Fig. 3.2).

Putting
$$y' = u, \tag{3.3.1}$$

the integral to be minimized is expressed in the form

$$C = \int_{x_0}^{x_1} F(x, y, u)\, dx \tag{3.3.2}$$

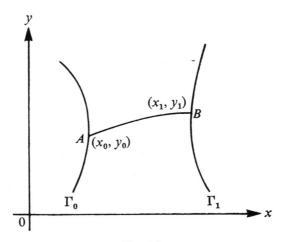

Fig. 3.2

ADDITIONAL CONSTRAINTS

and the problem has been reduced to a simple case of the optimization problem already solved. Clearly, the independent variable x is here playing the role formerly played by the time t, y is the state variable and u is the control variable. If $P_0(x, y) = 0$ and $P_1(x, y) = 0$ are the equations of the curves Γ_0, Γ_1, end constraints of the form

$$P_0(x_0, y_0) = 0, \qquad P_1(x_1, y_1) = 0 \tag{3.3.3}$$

are applicable.

The Hamiltonian for the problem is

$$H = F(x, y, u) + \lambda u \tag{3.3.4}$$

and the Hamilton equation for the multiplier λ is then derived in the form

$$\lambda' = -\frac{\partial H}{\partial y} = -\frac{\partial F}{\partial y}. \tag{3.3.5}$$

The optimal control law takes the form

$$\frac{\partial H}{\partial u} = \frac{\partial F}{\partial u} + \lambda = 0. \tag{3.3.6}$$

Elimination of λ between the last two equations leads to the equation

$$\frac{d}{dx}\left(\frac{\partial F}{\partial u}\right) = \frac{\partial F}{\partial y}. \tag{3.3.7}$$

Replacement of u by y' (equation (3.3.1)) finally yields *Euler's equation* for the problem, namely,

$$\frac{d}{dx}\left(\frac{\partial F}{\partial y'}\right) = \frac{\partial F}{\partial y}. \tag{3.3.8}$$

This is a second order differential equation for the unknown function $y(x)$.

The Hamiltonian for the constraint upon (x_1, y_1) is

$$J = \nu_1 P_1(x_1, y_1). \tag{3.3.9}$$

The end conditions (3.1.30) can now be written down:

$$\lambda_1 = \nu_1 \frac{\partial P_1}{\partial y_1}, \qquad H_1 = -\nu_1 \frac{\partial P_1}{\partial x_1}. \tag{3.3.10}$$

Eliminating the multiplier ν_1 and using equations (3.3.1), (3.3.4), (3.3.6), we deduce the condition

$$\left(\frac{\partial F}{\partial y'}\right)_1 \frac{\partial P_1}{\partial x_1} + \left(y' \frac{\partial F}{\partial y'} - F\right)_1 \frac{\partial P_1}{\partial y_1} = 0. \qquad (3.3.11)$$

Using the terminology of the calculus of variations, this is the *transversality condition* to be satisfied at the end point B.

Although in section 3.1, it has been assumed that the initial state of the system and the initial time are prescribed, it is easily seen how the argument must be amended to allow for these quantities to be variable and subject to given constraints. In these circumstances, conditions of the form (3.1.30) must be satisfied at $t = t^0$. For the problem being studied in this section, these conditions are shown to be equivalent to a transversality condition

$$\left(\frac{\partial F}{\partial y'}\right)_0 \frac{\partial P_0}{\partial x_0} + \left(y' \frac{\partial F}{\partial y'} - F\right)_0 \frac{\partial P_0}{\partial y_0} = 0 \qquad (3.3.12)$$

as explained above.

In the special case when F is independent of x, Euler's equation (3.3.8) must possess an integral $H = $ constant. This integral is equivalent to the equation

$$F - y' \frac{\partial F}{\partial y'} = \text{constant}. \qquad (3.3.13)$$

If y is absent from F, Euler's equation reduces to the form

$$\frac{d}{dx}\left(\frac{\partial F}{\partial y'}\right) = 0 \qquad (3.3.14)$$

and there is a first integral

$$\frac{\partial F}{\partial y'} = \text{constant}. \qquad (3.3.15)$$

PROBLEM 21. A uniform elastic string has an unstretched length l, a modulus of elasticity λ and a total weight wl. A point on it which is at a distance x from the end O when the string is unstretched, lies at a distance $x + y$ from O when the string hangs in equilibrium from this point. By minimizing the potential energy of the string, prove that

$$y = \frac{w}{2\lambda} x(2l - x).$$

ADDITIONAL CONSTRAINTS

Solution: Let P, P' be points on the string whose distances from O, when the string is unstretched, are $x, x + dx$ respectively. When the string hangs in equilibrium, their distances from O will be $x + y$ and $x + dx + y + dy$. Thus, the element dx of the string suffers an extension dy and its internal potential energy is equal to

$$\frac{\lambda}{2\,dx}\,dy^2 = \tfrac{1}{2}\lambda y'^2\,dx.$$

However, since this element descends a distance y, its gravitational potential energy decreases by $wy\,dx$. Thus, the net potential energy of the string can be taken to be given by

$$W = \tfrac{1}{2}\lambda \int_0^l (y'^2 - cy)\,dx,$$

where $c = 2w/\lambda$.

W has now to be minimized by choice of $y(x)$, subject to the end condition $y = 0$ at $x = 0$. The value of y at $x = l$ is not predetermined and it follows that the point (x_1, y_1) is only known to lie on the line $x_1 = l$; hence, $P_1(x_1, y_1) = x_1 - l$.

Euler's equation for the problem is

$$\frac{d}{dx}(2y') = -c.$$

Integrating twice, it is found that

$$y = -\tfrac{1}{4}cx^2 + Ax,$$

since $y = 0$ for $x = 0$.

Since $\partial P_1/\partial x_1 = 1$, $\partial P_1/\partial y_1 = 0$, the transversality condition (3.3.11) requires that $\partial F/\partial y' = \lambda y' = 0$ at $x = l$. Hence, $A = cl/2$ and

$$y = \tfrac{1}{4}cx(2l - x).$$

This is equivalent to the result stated. ●

3.4 Non-differential constraints

Consider, again, the problem posed in section 3.1, supposing that the control and state vectors are now subject to a constraint of the form

$$\phi(x, u, t) = 0, \tag{3.4.1}$$

in addition to the differential constraints provided by the state equations (2.1.2). It will be assumed that ϕ is explicitly dependent upon at least one control variable u_r; in this case, equation (3.4.1) can be solved for one of these control variables, thus enabling us to eliminate this variable from the problem. Then, the number of independent control variables is $(N-1)$ and the problem has been reduced to one of the type already studied. However, it may be more convenient to avoid such an elimination and to modify the argument of section 3.1 to take account of the dependence between the components of u. This can be done as follows:

The varied state and control vectors $y(t, \epsilon)$, $v(t, \epsilon)$ must now satisfy the condition
$$\phi[y(t, \epsilon), v(t, \epsilon), t] = 0 \tag{3.4.2}$$
identically in t and the ϵ_n. Partially differentiating with respect to ϵ_n and then setting $\epsilon = 0$, we find therefore that
$$\frac{\partial \phi}{\partial x_i} \eta_{ni} + \frac{\partial \phi}{\partial u_r} \xi_{nr} = 0. \tag{3.4.3}$$
Hence, the variations ξ_{nr} are no longer arbitrary functions of t, but are related by this equation.

The adjoint multipliers $\lambda_i(t)$ will now be required to satisfy equations
$$\dot{\lambda}_i = -\lambda_j \frac{\partial f_j}{\partial x_i} - \frac{\partial g}{\partial x_i} - \mu \frac{\partial \phi}{\partial x_i}, \tag{3.4.4}$$
instead of the equations (3.1.17); at this stage, $\mu(t)$ is an arbitrary function of t, but it will become determinate later. The amended form of equation (3.1.18) is now found to be
$$\frac{d}{dt}(\lambda_i \eta_{ni}) = \lambda_i \frac{\partial f_i}{\partial u_r} \xi_{nr} - \eta_{ni}\left(\frac{\partial g}{\partial x_i} + \mu \frac{\partial \phi}{\partial x_i}\right). \tag{3.4.5}$$
Appeal to equation (3.4.3) shows that this equation is equivalent to
$$\frac{d}{dt}(\lambda_i \eta_{ni}) = \left(\mu \frac{\partial \phi}{\partial u_r} + \lambda_i \frac{\partial f_i}{\partial u_r}\right)\xi_{nr} - \eta_{ni} \frac{\partial g}{\partial x_i} \tag{3.4.6}$$
and this permits our writing equation (3.1.14) in the form (3.1.19), where the Hamiltonian H must now be taken to be given by
$$H = g + \lambda_i f_i + \mu \phi. \tag{3.4.7}$$

ADDITIONAL CONSTRAINTS 83

The equation (3.1.24), namely

$$\int_{t_0}^{t_1} \frac{\partial H}{\partial u_r} \xi_{nr} \, dt = 0, \tag{3.4.8}$$

now follows by the argument set out in section 3.1, but the conclusion that $\partial H/\partial u_r$ vanishes cannot be reached immediately, since the ξ_{nr} are not independent functions. Instead, we first specify $\mu(t)$ by requiring that

$$\partial H/\partial u_1 = 0. \tag{3.4.9}$$

(Note: We are assuming $\partial \phi/\partial u_1 \neq 0$.) The condition (3.4.8) then becomes

$$\sum_{r=2}^{N} \int_{t_0}^{t_1} \frac{\partial H}{\partial u_r} \xi_{nr} \, dt = 0. \tag{3.4.10}$$

Now, provided $\partial \phi/\partial u_1 \neq 0$, the functions $\xi_{nr}(t)$ ($r = 2, 3, \ldots, N$) can be chosen arbitrarily and the constraint (3.4.3) satisfied by appropriate choice of $\xi_{n1}(t)$. It then follows from the condition (3.4.10) that

$$\frac{\partial H}{\partial u_r} = 0 \tag{3.4.11}$$

for $r = 2, 3, \ldots, N$. Taken in conjunction with equation (3.4.9), this means that the condition (3.4.11) is valid for all values of r (as deduced previously).

Since $\phi^1 = 0$, the end condition (3.1.28) remains valid.

If $\partial \phi/\partial u_1$ vanishes identically, it must be possible to find an alternative $\partial \phi/\partial u_R$ which does not vanish (otherwise, ϕ will be independent of the control vector) and the above argument can then be validated by fixing μ to satisfy $\partial H/\partial u_R = 0$.

After noting that the adjoint equations (3.4.4) are expressible in the Hamiltonian form $\dot{\lambda}_i = -\partial H/\partial x_i$, we can summarise our findings by stating that the conditions (3.1.29), (3.1.30) remain valid provided that the Hamiltonian is taken in the form (3.4.7).

If more than one constraint of the type (3.4.1) is operative, the Hamiltonian (3.4.7) must be augmented by the addition of further terms involving more multipliers $\mu_1, \mu_2, \ldots$.

In the special case when the constraint (3.4.1) is independent of u, i.e. takes the form

$$\phi(x, t) = 0, \tag{3.4.12}$$

by differentiating with respect to t, we derive the condition

$$\frac{d\phi}{dt} = \frac{\partial \phi}{\partial x_i} \dot{x}_i + \frac{\partial \phi}{\partial t} = 0. \tag{3.4.13}$$

Using the state equations, this is seen to be equivalent to the constraint

$$\phi_1(x, u, t) = \frac{\partial \phi}{\partial x_i} f_i + \frac{\partial \phi}{\partial t} = 0. \tag{3.4.14}$$

Provided ϕ_1 depends upon u, this latter constraint can now be allowed for as already explained; note, however, that equation (3.4.12) cannot then be completely ignored, since it leads to the additional terminal constraints

$$\phi(x^0, t^0) = 0, \qquad \phi(x^1, t^1) = 0. \tag{3.4.15}$$

Either one of these, coupled with equation (3.4.14) implies the other. It follows that it is necessary to adjoin only one of these end conditions to the set of constraints (3.1.1).

If ϕ_1 is independent of u, a further differentiation with respect to t will be necessary and one of the new terminal constraints

$$\phi_1(x^0, t^0) = \phi_1(x^1, t^1) = 0 \tag{3.4.16}$$

must be adjoined to (3.1.1). This process of differentiation continues, until a constraint of the form (3.4.1) is generated.

PROBLEM 22. q_i ($i = 1, 2, \ldots, M$) are generalized coordinates for a certain mechanical system and $L(\dot{q}, q, t)$ is its Lagrangian. The coordinates are not independent but are related by the equations

$$\phi_r(q_i) = 0,$$

where $r = 1, 2, \ldots, N$ ($N < M$). Employ Hamilton's principle to derive its equations of motion.

Solution: Introducing new variables u_i by means of equations

$$\dot{q}_i = u_i,$$

we shall regard the u_i as control variables and the q_i as state variables.

According to Hamilton's principle, the motion of the system between a specified configuration $q = q^0$ at $t = t^0$ and another given

ADDITIONAL CONSTRAINTS

configuration $q = q^1$ at $t = t^1$ will be such that the integral

$$C = \int_{t^0}^{t^1} L(u, q, t)\, dt$$

is stationary with respect to all small deviations from this motion. Thus, our problem is to choose the control vector $u(t)$ so that the terminal constraints on the state vector $q(t)$ are satisfied and the system behaviour is optimized with respect to the performance index C; the control is also to be such that the state vector satisfies the constraints $\phi_r(q) = 0$.

Since these last constraints are independent of u, they must first be differentiated with respect to t to yield new constraints

$$\frac{\partial \phi_r}{\partial q_i} u_i = 0.$$

The Hamiltonian can now be constructed in the form

$$H = L + \lambda_i u_i + \mu_r \frac{\partial \phi_r}{\partial q_i} u_i,$$

with multipliers λ_i, μ_r. The equations determining the optimal behaviour of the system are then found to be

$$\dot{\lambda}_i = -\frac{\partial H}{\partial q_i} = -\frac{\partial L}{\partial q_i} - \mu_r \frac{\partial^2 \phi_r}{\partial q_i \partial q_j} u_j,$$

$$\frac{\partial H}{\partial u_i} = \frac{\partial L}{\partial u_i} + \lambda_i + \mu_r \frac{\partial \phi_r}{\partial q_i} = 0.$$

Since the state variables take known values at both terminals the λ_i are subject to no end conditions.

Eliminating λ_i between the last pair of equations, we find

$$\frac{d}{dt}\left(\frac{\partial L}{\partial u_i}\right) + \frac{d}{dt}\left(\mu_r \frac{\partial \phi_r}{\partial q_i}\right) = \frac{\partial L}{\partial q_i} + \mu_r \frac{\partial^2 \phi_r}{\partial q_i \partial q_j} u_j.$$

But

$$\frac{d}{dt}\left(\mu_r \frac{\partial \phi_r}{\partial q_i}\right) = \dot{\mu}_r \frac{\partial \phi_r}{\partial q_i} + \mu_r \frac{\partial^2 \phi_r}{\partial q_i \partial q_j} \dot{q}_j.$$

Since $u_i = \dot{q}_i$, it now follows that

$$\frac{d}{dt}\left(\frac{\partial L}{\partial \dot{q}_i}\right) - \frac{\partial L}{\partial q_i} = -\dot{\mu}_r \frac{\partial \phi_r}{\partial q_i}.$$

Putting $-\dot{\mu}_r = \nu_r$, the system of equations of motion is reduced to its final form, namely

$$\frac{d}{dt}\left(\frac{\partial L}{\partial \dot{q}_i}\right) - \frac{\partial L}{\partial q_i} = \nu_r \frac{\partial \phi_r}{\partial q_i}.$$

These M equations, together with the N constraints $\phi_r = 0$, are sufficient to determine the M coordinates q_i and the N multipliers ν_r.●

3.5 Integral constraints

Constraints upon a system expressed in integral form are sometimes encountered. Thus, it might be a requirement that the system behaviour is to be such that the integral

$$\int_{t^0}^{t^1} \phi(x, u, t)\, dt, \tag{3.5.1}$$

where ϕ is a given function, takes a prescribed value a.

This type of problem can be reduced to a problem already studied by introducing a further state variable x_{M+1} satisfying the equation

$$\dot{x}_{M+1} = \phi(x, u, t) \tag{3.5.2}$$

and the initial condition $x_{M+1} = 0$ at $t = t^0$. Then, the integral constraint can be written as a terminal constraint upon x_{M+1}, namely

$$x_{M+1} = a, \qquad t = t^1. \tag{3.5.3}$$

Assuming that, apart from this integral constraint, the problem is as stated in section 3.1, the Hamiltonian is now taken in the form

$$H = g + \lambda_i f_i + \mu \phi, \tag{3.5.4}$$

where μ is an additional multiplier. Hamilton's equations are then written down in the usual way, that for the multiplier μ being

$$\dot{\mu} = -\frac{\partial H}{\partial x_{M+1}} = 0. \tag{3.5.5}$$

It follows that μ is a constant, its value being determined ultimately by the end condition (3.5.3).

PROBLEM 23. In the 4-terminal network illustrated in Fig. 2.1 (Problem 11), take $R = \tfrac{1}{3}$, $S = 1$, $C = 1$. If, initially, the capacitor is

ADDITIONAL CONSTRAINTS

uncharged, calculate the optimal control $u(t)$ to be applied across the input terminals, if the potential across the output terminals is to be raised to the value $x = 12$ in the minimum time and the energy supplied by the controller is to be 240. Show that the minimum time is $\frac{1}{2} \log 3$.

Solution: Employing equations already derived in the solution to Problem 11, it can be shown that the state equation for the network is

$$\dot{x} = 3u - 4x. \tag{3.5.6}$$

Denoting the initial and final instants by $t = 0$, T, respectively, the cost to be minimized is

$$C = \int_0^T dt.$$

The potential drop across the resistor R is $(u - x)$ and the current taken from the controller is accordingly $3(u - x)$; thus, the energy supplied by this device is given by

$$3 \int_0^T u(u - x)\, dt.$$

This is to take the value 240, leading to the integral constraint

$$\int_0^T u(u - x)\, dt = 80. \tag{3.5.7}$$

The Hamiltonian can now be constructed in the form

$$H = 1 + \lambda(3u - 4x) + \mu u(u - x) \tag{3.5.8}$$

and the equations

$$\dot{\lambda} = -\frac{\partial H}{\partial x} = 4\lambda + \mu u, \tag{3.5.9}$$

$$\frac{\partial H}{\partial u} = 3\lambda + 2\mu u - \mu x = 0, \tag{3.5.10}$$

follow immediately. Elimination of the variables λ, u between equations (3.5.6), (3.5.9), (3.5.10), yields the equation

$$\ddot{x} - 4x = 0. \tag{3.5.11}$$

Since $x = 0$ at $t = 0$, the required solution of this equation takes the form

$$x = A \sinh 2t. \tag{3.5.12}$$

The end condition $x = 12$, $t = T$, then requires that

$$A \sinh 2T = 12. \tag{3.5.13}$$

Equation (3.5.6) now shows that

$$u = \tfrac{2}{3}A(\cosh 2t + 2 \sinh 2t) \tag{3.5.14}$$

and the constraint (3.5.7) will thus be found to demand that

$$\tfrac{1}{18}A^2(4 \sinh 2T \cosh 2T + 5 \sinh^2 2T) = 80. \tag{3.5.15}$$

Equations (3.5.13), (3.5.15), can now be solved for A and T; the results are $A = 9$, $\sinh 2T = \tfrac{4}{3}$. The last equation implies that $e^{2T} = 3$ and, hence, that $T = \tfrac{1}{2} \log 3$.

Thus, the optimal control is determined by

$$u = 6 \cosh 2t + 12 \sinh 2t$$

and the potential produced across the output terminals is

$$x = 9 \sinh 2t.$$

Since T is variable, $H = 0$ must be an integral of the Hamilton equations; it will be found that this requires that $\mu = \tfrac{1}{36}$. ●

3.6 Discontinuous controls

Consider the optimal control problem formulated in section 3.1 in the special case when the end constraints (3.1.1) take the form

$$x_i^1 = P_i(t^1), \tag{3.6.1}$$

where $i = 1, 2, \ldots, M$ and the P_i are given functions possessing continuous first derivatives. These constraints may be described as requiring that the point representing the system's final state must lie on a certain curve Γ in state space; equations (3.6.1) are parametric equations for Γ. If the value of t^1 is predetermined, the final state will be completely fixed and the optimal behaviour and control can be found by solving the equations (3.1.29) subject to the end conditions satisfied by the state vector at $t = t^0$, t^1 (the λ_i and H are subject to no end conditions, in this case). This solution will depend upon

ADDITIONAL CONSTRAINTS

the predetermined value of t^1 and will accordingly take the form

$$x_i = x_i(t, t^1), \qquad u_r = u_r(t, t^1). \tag{3.6.2}$$

We shall now study the manner in which the optimal cost depends upon the parameter t^1.

We have

$$C = G(x^1, t^1) + \int_{t^0}^{t^1} g(x, u, t)\, dt, \tag{3.6.3}$$

where the vectors x, u are given by equations (3.6.2) and x^1 is found from equations (3.6.1). Differentiating C with respect to the parameter t^1, we find

$$\frac{dC}{dt^1} = \frac{dG}{dt^1} + g(x^1, u^1, t^1) + \int_{t^0}^{t^1} \left(\frac{\partial g}{\partial x_i} \frac{\partial x_i}{\partial t^1} + \frac{\partial g}{\partial u_r} \frac{\partial u_r}{\partial t^1} \right) dt, \tag{3.6.4}$$

where

$$\frac{dG}{dt^1} = \frac{\partial G}{\partial x_i^1} \frac{dP_i}{\partial t^1} + \frac{\partial G}{\partial t^1}. \tag{3.6.5}$$

Since equations (3.1.30) are satisfied along an optimal trajectory, it follows that

$$\frac{\partial g}{\partial x_i} = -\dot\lambda_i - \lambda_j \frac{\partial f_j}{\partial x_i}, \tag{3.6.6}$$

$$\frac{\partial g}{\partial u_r} = -\lambda_j \frac{\partial f_j}{\partial u_r}. \tag{3.6.7}$$

Thus, equation (3.6.4) can be written in the form

$$\frac{dC}{dt^1} = \frac{dG}{dt^1} + g^1 - \int_{t^0}^{t^1} \left[\left(\dot\lambda_i + \lambda_j \frac{\partial f_j}{\partial x_i} \right) \frac{\partial x_i}{\partial t^1} + \lambda_j \frac{\partial f_j}{\partial u_r} \frac{\partial u_r}{\partial t^1} \right] dt. \tag{3.6.8}$$

But, the state equations must also be satisfied by the optimal solution (3.6.2), i.e.

$$\dot x_j(t, t^1) = f_j[x(t, t^1), u(t, t^1), t] \tag{3.6.9}$$

is an identity in the variables t and t^1. Partially differentiating the identity with respect to t^1, we obtain the result

$$\frac{\partial^2 x_j}{\partial t\, \partial t^1} = \frac{\partial f_j}{\partial x_i} \frac{\partial x_i}{\partial t^1} + \frac{\partial f_j}{\partial u_r} \frac{\partial u_r}{\partial t^1}. \tag{3.6.10}$$

Equation (3.6.8) can now be expressed in the form

$$\frac{dC}{dt^1} = \frac{dG}{dt^1} + g^1 - \int_{t^0}^{t^1}\left(\lambda_i \frac{\partial x_i}{\partial t^1} + \lambda_j \frac{\partial^2 x_j}{\partial t\, \partial t^1}\right) dt$$

$$= \frac{dG}{dt^1} + g^1 - \left|\lambda_i \frac{\partial x_i}{\partial t^1}\right|_{t^0}^{t^1}$$

$$= \frac{dG}{dt^1} + g^1 - \lambda_i^1\left(\frac{\partial x_i}{\partial t^1}\right)_{t=t^1} \quad (3.6.11)$$

since the initial conditions $x_i(t^0, t^1) = x_i^0$ imply that $\partial x_i/\partial t^1$ all vanish at $t = t^0$.

The constraints (3.6.1) require that

$$x_i(t^1, t^1) = P_i(t^1). \quad (3.6.12)$$

Hence, differentiating with respect to t^1, we find that

$$\left(\frac{\partial x_i}{\partial t}\right)_{t=t^1} + \left(\frac{\partial x_i}{\partial t^1}\right)_{t=t^1} = \frac{dP_i}{dt^1}. \quad (3.6.13)$$

Since $\partial x_i/\partial t = f_i$, it follows that

$$\left(\frac{\partial x_i}{\partial t^1}\right)_{t=t^1} = \frac{dP_i}{dt^1} - f_i^1, \quad (3.6.14)$$

thus permitting us to write equation (3.6.11) in the form

$$\frac{dC}{dt^1} = \frac{dG}{dt^1} + H^1 - \lambda_i^1 \frac{dP_i}{dt^1} \quad (3.6.15)$$

($H = g + \lambda_i f_i$).

By minor amendment of the argument it can be proved that the result we have just found remains valid (with suitably amended Hamiltonian) for the problems studied in sections 3.4 and 3.5.

We shall employ this last result to establish necessary conditions to be satisfied at any discontinuity in the optimal control function $u(t)$. Suppose such a discontinuity exists at the instant $t = t^*$, where $t^0 < t^* < t^1$. We shall embed the optimal trajectory for the system in a family of admissible trajectories as follows: Let the equations

$$x_i = P_i(\tau) \quad (3.6.16)$$

define a curve Γ in control space, passing through the point x_i^* which represents the optimal state of the system at the instant $t = t^*$; let

ADDITIONAL CONSTRAINTS

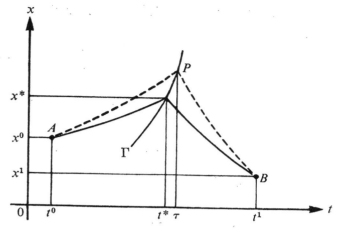

Fig. 3.3

$\tau = t^*$ at this point on the curve. Construct the optimal trajectory which transfers the system from the initial state $x = x^0$ at $t = t^0$ to the final state $x = P(\tau)$ at $t = \tau$; construct, also, the optimal trajectory connecting the states $x = P(\tau)$ at $t = \tau$ and $x = x^1$ at $t = t^1$. Variation of τ over a range of values including the value t^* now defines a family Φ of trajectories within which the optimal trajectory is embedded (Fig. 3.3).

The cost associated with an arbitrary member APB (Fig. 3.3) of Φ is a function of the parameter τ and is given by

$$C(\tau) = C^-(\tau) + C^+(\tau) + G(x^1, t^1), \tag{3.6.17}$$

where

$$C^- = \int_{AP} g \, dt, \qquad C^+ = \int_{PB} g \, dt. \tag{3.6.18}$$

Making use of the result (3.6.15) (with $G = 0$), it follows that

$$\left.\begin{aligned}\frac{dC^-}{d\tau} &= H^- - \lambda_i^- \frac{dP_i}{d\tau}, \\ \frac{dC^+}{d\tau} &= -H^+ + \lambda_i^+ \frac{dP_i}{d\tau},\end{aligned}\right\} \tag{3.6.19}$$

H^-, λ_i^- being calculated at P on the trajectory AP and H^+, λ_i^+ being calculated at P on the trajectory PB. (Note: the sign change in the

92 ANALYTICAL METHODS OF OPTIMIZATION

second equation is a consequence of P being the initial point on the trajectory PB.) Hence,

$$\frac{dC}{d\tau} = H^- - H^+ + (\lambda_i^+ - \lambda_i^-)\frac{dP_i}{d\tau}. \qquad (3.6.20)$$

Since C must be stationary for $\tau = t^*$, $dC/d\tau$ must vanish for this value of τ. But the derivatives $dP_i/d\tau$ can be given arbitrary values at $\tau = t^*$ by choice of Γ. We conclude that

$$H(t^* - 0) = H(t^* + 0), \qquad \lambda_i(t^* - 0) = \lambda_i(t^* + 0). \quad (3.6.21)$$

These are necessary conditions to be satisfied at any discontinuity in the control u.

PROBLEM 24. A system is governed by the state equation $\dot{x} = u$ and the cost to be minimized is

$$C = \int_0^1 (u^2 - 1)^2 \, dt.$$

The system is to be transferred from an initial state $x = 0$ at $t = 0$ to a final state $x = X$ at $t = 1$. Show that the control is discontinuous if $|X| < 1$ and is continuous otherwise.

Solution: The Hamiltonian is

$$H = (u^2 - 1)^2 + \lambda u.$$

Thus, $\dot{\lambda} = 0$; i.e. λ is constant.

For an optimal control

$$\frac{\partial H}{\partial u} = 4u(u^2 - 1) + \lambda = 0$$

and, hence, u is constant.

At any discontinuity in u, it is necessary that λ be continuous and

$$(u_-^2 - 1)^2 + \lambda u_- = (u_+^2 - 1)^2 + \lambda u_+, \qquad (3.6.22)$$

$$\lambda = -4u_-(u_-^2 - 1) = -4u_+(u_+^2 - 1), \qquad (3.6.23)$$

where u_-, u_+ are the values taken by u before and after the discontinuity, respectively. Substituting for λ into equation (3.6.22), this

ADDITIONAL CONSTRAINTS 93

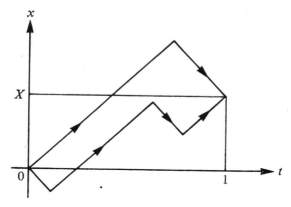

Fig. 3.4

can be written
$$(u_-^2 - u_+^2)(u_-^2 + u_+^2 - \tfrac{2}{3}) = 0.$$
Since $u_- \neq u_+$ at a discontinuity, this implies that
$$u_+ = -u_- \quad \text{or} \quad u_-^2 + u_+^2 = \tfrac{2}{3}. \tag{3.6.24}$$
Equation (3.6.23) leads to the condition
$$u_-^2 + u_- u_+ + u_+^2 = 1.$$
Taking this condition with the second of equations (3.6.24), we calculate that $u_- = u_+ = \pm 1/\sqrt{3}$, i.e. there is no discontinuity. Rejecting this possibility and accepting that $u_+ = -u_-$, we find that $u_+ = -u_- = \pm 1$.

It has been shown, therefore, that there are two possibilities for the optimal control law. Either (i) u maintains the same constant value from $t = 0$ to $t = 1$, or (ii) u oscillates between the values $+1$ and -1 during the interval $0 \leqslant t \leqslant 1$.

If $|X| > 1$, it is clear that the magnitude of the gradient of the optimal trajectory in the xt-plane must exceed unity for some values of t. Thus, the discontinuous control is ruled out as a possibility in this case. In such a case, to satisfy the end conditions it is necessary to take $u = X$, $x = Xt$ and, hence, $C_{\text{opt}} = (X^2 - 1)^2$.

If, however, $|X| < 1$, the discontinuous mode of control is a possibility to be considered and, in fact, any number of modes of this type exist, satisfying the end conditions; two are indicated in Fig. 3.4.

Since $u^2 = 1$ along all such trajectories, C_{opt} vanishes in every such case. But C can never be negative, so that such modes of control are certainly optimal. ●

3.7 Linear systems

We shall now reconsider the problem, already studied in section 2.7, of optimizing the performance of a linear system subject to the quadratic performance index given by equation (2.7.1). However, the final state x^1 will no longer be supposed free, but will be required to satisfy the end constraints

$$x_m^1 = X_m, \qquad (3.7.1)$$

where $m = 1, 2, \ldots, Q$ and the X_m are given quantities. Since only the components x_p^1 ($p = Q + 1, Q + 2, \ldots, M$) of x^1 can vary in value, the cost function reduces to the form

$$C = \tfrac{1}{2}s_{pq}x_p^1 x_q^1 + \tfrac{1}{2}\int_{t_0}^{t^1}(p_{ij}x_i x_j + 2q_{ir}x_i u_r + r_{rs}u_r u_s)dt \qquad (3.7.2)$$

where p, q range over the values $Q + 1, \ldots, M$.

The Hamiltonian for the problem can now be written down as at equation (2.7.2) and Hamilton's equations reduce to the form (2.7.10) as previously. By equations (3.1.33), the multipliers are subject to the end conditions

$$\lambda_p^1 = s_{pq}x_q^1. \qquad (3.7.3)$$

Let S be the square matrix of order $(M - Q)$ whose elements are s_{pq} and let I_Q be the unit matrix of order Q. Constructing an $M \times M$ matrix T and an $M \times Q$ matrix J thus,

$$T = \begin{bmatrix} 0 & 0 \\ 0 & S \end{bmatrix}, \quad J = \begin{bmatrix} I_Q \\ 0 \end{bmatrix}, \qquad (3.7.4)$$

we can write

$$\lambda^1 = Tx^1 + J\alpha, \qquad (3.7.5)$$

where α is the column matrix with elements $\alpha_m = \lambda_m^1$ ($m = 1, 2, \ldots, Q$), i.e. the components of λ not subject to the end constraint (3.7.3).

Hamilton's equations (2.7.10) are now solved in the form (2.7.15) as before. Substituting for λ^1 from equation (3.7.5), these equations

ADDITIONAL CONSTRAINTS

then yield
$$x(t) = (\phi_1 + \phi_2 T)x^1 + \phi_2 J\alpha, \\ \lambda(t) = (\phi_3 + \phi_4 T)x^1 + \phi_4 J\alpha.$$ (3.7.6)

Eliminating x^1 between this pair of equations, we find that
$$\lambda = Kx + U\alpha,$$ (3.7.7)
where
$$K(t, t^1) = (\phi_3 + \phi_4 T)(\phi_1 + \phi_2 T)^{-1},$$ (3.7.8)
$$U(t, t^1) = [\phi_4 - (\phi_3 + \phi_4 T)(\phi_1 + \phi_2 T)^{-1}\phi_2]J.$$ (3.7.9)

Clearly, K is an $M \times M$ matrix and U is an $M \times Q$ matrix. If we now substitute for λ from equation (3.7.7) into the Hamilton equations (2.7.10), treating α as a constant vector, the following equations are obtained:

$$\dot{x} = (A - BR^{-1}Q^T - BR^{-1}B^TK)x - BR^{-1}B^TU\alpha,$$ (3.7.10)
$$K\dot{x} + \dot{K}x + \dot{U}\alpha = (QR^{-1}Q^T - P + QR^{-1}B^TK - A^TK)x \\ + (QR^{-1}B^T - A^T)U\alpha.$$ (3.7.11)

Eliminating $\dot{x}$, we deduce that

$$[\dot{K} + P + KA + A^TK - (KB + Q)R^{-1}(B^TK + Q^T)]x \\ + [\dot{U} + (A^T - QR^{-1}B^T - KBR^{-1}B^T)U]\alpha = 0.$$ (3.7.12)

Since, in establishing this equation, no use has been made of the initial conditions $x = x^0$ or the end conditions (3.7.1), the vectors x, α can be given arbitrary values. Thus, it follows from the last equation that

$$\dot{K} = (KB + Q)R^{-1}(B^TK + Q^T) - KA - A^TK - P,$$ (3.7.13)
$$\dot{U} = (KBR^{-1}B^T + QR^{-1}B^T - A^T)U.$$ (3.7.14)

The first of these equations is identical with the Riccati equation (2.7.22). Since, at $t = t^1$, $\phi_1 = \phi_4 = I$ and $\phi_2 = \phi_3 = 0$, equation (3.7.8) shows that equation (3.7.13) is to be solved under the end condition
$$K(t^1, t^1) = T.$$ (3.7.15)

Then, on account of the symmetry of T, we deduce as before that the matrix K is symmetric for all values of t.

Equation (3.7.14) is a linear equation for U and, by equation (3.7.9), is to be solved under the end condition

$$U(t^1, t^1) = J. \tag{3.7.16}$$

Having calculated K and U, equation (3.7.10) can be integrated forward from the the given state x^0 at $t = t^0$ to the final state x^1. The vector α is still unknown and has to be chosen in such a way that x^1 satisfies the end conditions (3.7.1). The proper choice for α can conveniently be calculated thus:

We regard α as a constant control vector and equation (3.7.10) as the state equation of a linear system. This equation will then possess a solution of the form (2.2.20), i.e.

$$x(t) = V(t, t^0)x^0 + W(t, t^0)\alpha, \tag{3.7.17}$$

where V is an $M \times M$ matrix and W is an $M \times Q$ matrix. V and W satisfy the conditions

$$V(t^0, t^0) = I_M, \qquad W(t^0, t^0) = 0. \tag{3.7.18}$$

The matrix equation (3.7.17) is equivalent to a set of M ordinary equations for $x_i(t)$, $i = 1, 2, \ldots, M$. By deleting the last $(M - Q)$ rows of V and W to give new matrices $Y(Q \times M)$, $Z(Q \times Q)$, we shall retain only the first Q of these equations; thus,

$$\bar{x}(t) = Y(t, t^0)x^0 + Z(t, t^0)\alpha, \tag{3.7.19}$$

where $\bar{x}(t) = [x_1, x_2, \ldots, x_Q]^T$. Replacing t by t^1 and t^0 by t, we next deduce that

$$X = Y(t^1, t)x(t) + Z(t^1, t)\alpha, \tag{3.7.20}$$

where the Q elements of the vector X are the quantities X_m given in the end conditions (3.7.1). Differentiating equation (3.7.20) with respect to t and substituting for $\dot{x}$ from equation (3.7.10), it follows that

$$0 = [\dot{Y} + Y(A - BR^{-1}Q^T - BR^{-1}B^TK)]x + (\dot{Z} - YBR^{-1}B^TU)\alpha. \tag{3.7.21}$$

This equation will be valid for arbitrary values of x^0 and X and, therefore, for arbitrary values of x and α. Hence

$$\dot{Y} = Y(BR^{-1}B^TK + BR^{-1}Q^T - A), \tag{3.7.22}$$

$$\dot{Z} = YBR^{-1}B^TU. \tag{3.7.23}$$

It follows from the first of equations (3.7.18) that the truncated matrix Y satisfies the end condition

$$Y(t^1, t^1) = [I_Q \quad 0] = J^T. \quad (3.7.24)$$

Also, by transposing equation (3.7.22), it will be found that Y^T satisfies the same equation as U, namely (3.7.14). We can now identify Y^T and U; thus,

$$Y(t^1, t) = U^T(t, t^1). \quad (3.7.25)$$

The equation (3.7.23) for Z accordingly reads

$$\dot{Z} = U^T B R^{-1} B^T U. \quad (3.7.26)$$

The end condition from which this equation is to be integrated is derived from the second of equations (3.7.18); it is

$$Z(t^1, t^1) = 0. \quad (3.7.27)$$

Equations (3.7.26), (3.7.27), imply that Z is symmetric.

Finally, putting $t = t^0$ in equation (3.7.20) and solving for α, we get

$$\alpha = Z^{-1}(t^1, t^0)[X - U^T(t^0, t^1)x^0]. \quad (3.7.28)$$

This fixes α in terms of the initial and final state vectors x^0, X.

The procedure, therefore, is first to integrate equations (3.7.13), (3.7.14), (3.7.26) backwards from the final instant $t = t^1$ for K, U and Z, making use of the end conditions (3.7.15), (3.7.16), (3.7.27). Equation (3.7.28) then yields the vector α and equation (3.7.10) can be integrated forwards from $t = t^0$ to generate the system's optimal behaviour. The optimal control follows from equations (2.7.9) and (3.7.7); thus, eliminating λ,

$$u = -R^{-1}[(Q^T + B^T K)x + B^T U\alpha]. \quad (3.7.29)$$

It has been tacitly assumed that t^1 is predetermined. If this is not the case, H must satisfy the end condition (3.1.33) which, in the case we are considering, takes the form

$$H^1 + \frac{ds_{pq}}{dt^1} x_p^1 x_q^1 = 0. \quad (3.7.30)$$

If a numerical mode of solution is being performed, it will generally be necessary first to choose a series of trial values for t^1 and to calculate the value of the left-hand member of equation (3.7.30) in each case; interpolation can then be used to derive a value of t^1 for which this member will vanish.

If, however, the matrix S is independent of t^1 and the linear system is autonomous so that the matrices A, B, P, Q, R are independent of t, a more direct procedure can be adopted. In this case, $H = 0$ will be a first integral of Hamilton's equations. Giving t^1 a purely nominal value (e.g. zero), the matrices K, U, Z can be calculated by backwards integration. After each step, the possibility that t^0 has been reached is tested as follows: α is calculated from equation (3.7.28) using the known value of x^0 and the possible values of λ^0, u^0 are then found from equations (3.7.7), (2.7.9); these values of x^0, λ^0, u^0 are substituted in the Hamiltonian and, if this vanishes, t^0 has been reached. As soon as the point in the range of integration corresponding to t^0 has been identified, the nominal value of t^1 can be corrected to the real value.

PROBLEM 25. The 4-terminal network described in Problem 11 is to be brought from the state $x = 0$ at $t = 0$ to the state $x = 1$ at $t = 1$ by a control $u(t)$ which minimizes the stated performance index. Calculate the optimal control and behaviour.

Solution: The state equation is $\dot{x} = -2x + u$ and the integrand of the cost integral is $x^2 + u^2/5$. The matrices of the general theory are accordingly all of the first order and are given by $A = -2$, $B = 1$, $S = 0$, $P = 2$, $Q = 0$, $R = \frac{2}{5}$, $T = 0$, $J = 1$. Equation (3.7.13) takes the form

$$\dot{K} = \tfrac{1}{2}(5K - 2)(K + 2)$$

and integrates under the condition $K = T = 0$ at $t = 1$ to give

$$K = \frac{2 - 2e^{6(t-1)}}{5 + e^{6(t-1)}}.$$

Equation (3.7.14) can now be written down thus,

$$\dot{U} = \frac{15 - 3e^{6(t-1)}}{5 + e^{6(t-1)}} U.$$

Integration under the condition $U = J = 1$ at $t = 1$ leads to the result

$$U = \frac{6e^{3(t-1)}}{5 + e^{6(t-1)}}.$$

ADDITIONAL CONSTRAINTS

Equation (3.7.26) for Z is now found in the form

$$\dot{Z} = \frac{90e^{6(t-1)}}{(5 + e^{6(t-1)})^2}.$$

Integrating with $Z = 0$ at $t = 1$, we find that

$$\dot{Z} = \frac{5e^{6(t-1)} - 5}{2e^{6(t-1)} + 10}.$$

Equation (3.7.28) (with $X = 1$, $x^0 = 0$, $t^0 = 0$) yields

$$\alpha = \frac{2 + 10e^6}{5 - 5e^6}$$

and equation (3.7.10) for the system's optimal behaviour therefore takes the form

$$\dot{x} + \frac{15 - 3e^{6(t-1)}}{5 + e^{6(t-1)}} x = \frac{6(1 + 5e^6)e^{3(t-1)}}{(e^6 - 1)(5 + e^{6(t-1)})}.$$

Multiplying this equation by the integrating factor $e^{3(t-1)}/[5 + e^{6(t-1)}]$ and integrating under the initial condition $x = 0$ at $t = 0$, it is found that

$$x = \frac{\sinh 3t}{\sinh 3}.$$

The optimal control is now derived from the state equation to be

$$u = (3 \cosh 3t + 2 \sinh 3t)/\sinh 3,$$

and the optimal cost is therefore $(3 \coth 3 + 2)/5$. ●

A simple formula for the optimal cost can be derived as follows: If $x(t)$ is the optimal state function, by use of equations (3.7.10), (3.7.13) we first show that

$$\frac{d}{dt}(x^T K x) = \dot{x}^T K x + x^T \dot{K} x + x^T K \dot{x}$$
$$= x^T(QR^{-1}Q^T - P - KBR^{-1}B^T K)x$$
$$- \alpha^T U^T BR^{-1}B^T K x - x^T KBR^{-1}B^T U\alpha. \quad (3.7.31)$$

(Recall that the matrices R and K are symmetric.)

The cost C (equation (3.7.2)) is next written in the matrix form

$$C = \tfrac{1}{2}(x^T T x)^1 + \tfrac{1}{2}\int_{t_0}^{t_1}(x^T P x + x^T Q u + u^T Q^T x + u^T R u)\, dt.$$

(3.7.32)

Then, by substituting for u from the optimal control law (3.7.29), the integrand of the cost integral can be reduced, using equation (3.7.31), thus:

$$x^T Px + x^T Qu + u^T Q^T x + u^T Ru$$
$$= -\frac{d}{dt}(x^T Kx) + \alpha^T U^T BR^{-1}B^T U\alpha. \quad (3.7.33)$$

It now follows by equation (3.7.26) that the integrand can be expressed in the form

$$\frac{d}{dt}(\alpha^T Z\alpha - x^T Kx). \quad (3.7.34)$$

Hence

$$C_{\text{opt}} = \tfrac{1}{2}(x^T Tx)^1 + \tfrac{1}{2}(\alpha^T Z\alpha - x^T Kx)^1 - \tfrac{1}{2}(\alpha^T Z\alpha - x^T Kx)^0. \quad (3.7.35)$$

But, at $t = t^1$, $K = T$ and $Z = 0$. Thus,

$$C_{\text{opt}} = \tfrac{1}{2}(x^T Kx - \alpha^T Z\alpha)^0. \quad (3.7.36)$$

In the special case of Problem 25, $x^0 = 0$ and, hence,

$$C_{\text{opt}} = -\tfrac{1}{2}(\alpha^T Z\alpha)^0 = \frac{5e^6 + 1}{5e^6 - 5}.$$

This agrees with the result already found.

Exercises 3

1. The motion of an oscillatory system is governed by the equation $\ddot{x} + x = u$, u being the applied force. The system is to be moved from its equilibrium state $x = \dot{x} = 0$ at $t = 0$ to the state $x = 1$, $\dot{x} = 0$ at $t = \tfrac{1}{2}\pi$, in such a way that the cost

$$C = \int_0^{\frac{1}{2}\pi} u^2 \, dt$$

is minimized. Calculate the optimal control $u(t)$ and deduce that $C_{\text{opt}} = 4\pi/(\pi^2 - 4)$.

2. A control system has a single state variable and a single control variable, x and u respectively, related by the equation $\dot{x} = x + u$.

The cost C to be minimized is given by

$$C = \int_{t_0}^{t_1} (x^2 + \tfrac{1}{3}u^2)\, dt.$$

If $t_0 = 0$, $t_1 = 1$ and the initial state is given by $x(0) = 1$, but the final state is not determined, calculate the optimal behaviour and the optimal control. Sketch the graph of x_{opt} against t and show that its final value is $4e^2/(e^4 + 3)$. If $t_0 = 0$, t_1 is undetermined and the initial and final states are given by $x(0) = 1$, $x(t_1) = e^{-1}$, again calculate the optimal behaviour and control and sketch the graph x_{opt}. Prove that $t_1 = \tfrac{1}{2}$ and the optimal cost is $1 - e^{-2}$.

3. An aeroplane has a fixed velocity V with respect to the air and the wind velocity w is uniform. The aeroplane is to fly a closed curve at a constant height taking a time T to make a circuit. If the area enclosed by the curve is to be a maximum, show that it is an ellipse of area

$$\frac{V^2 T^2}{4\pi}\left(1 - \frac{w^2}{V^2}\right)^{\frac{3}{2}}.$$

4. A second-order system has state equations

$$\dot{x} = y - x, \qquad \dot{y} = u - y,$$

u being the control quantity. The system is to be taken from the state $x = y = 0$ at $t = 0$ to a final state at $t = 1$ in which $x = 1$. The performance index to be minimized is

$$C = \int_0^1 u^2 \, dt.$$

Calculate the optimal control and show that $C_{\min} = 4e^2/(e^2 - 5)$.

5. A river is of width d and its velocity of flow at a distance x from one of the banks is given by $v = f(x)$. A ferry operates between two terminals, one on each bank, the speed of the ferry relative to the water being a constant V. Taking axes Ox, Oy through a terminal, show that, if the crossing time is to be minimized, the path of the ferry is given by

$$y = \frac{1}{V} \int_0^x \frac{v(v - \alpha) - V^2}{[(v - \alpha)^2 - V^2]^{\frac{1}{2}}}\, dx$$

where α is determined by the position of the other terminal. If both terminals lie on the x-axis and v is constant, prove that the minimum crossing time is $d/\sqrt{(V^2 - v^2)}$.

6. Employing the equations of motion for a rocket given in the solution to Problem 12 (equations (2.6.26)), calculate the optimal thrust direction programme for launching the rocket into an orbit about the earth. Specifically, assume that a, T are given constants, that the final horizontal velocity u_1 is prescribed and that the maximum height achieved after "all burnt" is to be maximized. It may be shown that the optimal thrust programme is determined by an equation of the form $\tan \theta = \alpha - \beta t$, where α, β are positive constants chosen so that the initial and final values of θ satisfy the equations

$$a(\sec \theta_0 - \sec \theta_1) = g \tan \theta_0$$

$$u_1(\tan \theta_0 - \tan \theta_1) = aT \log\left(\frac{\sec \theta_0 + \tan \theta_0}{\sec \theta_1 + \tan \theta_1}\right).$$

7. A system is governed by the equation $\dot{x} + x = u$ and its initial state is given by $x = 1$ at $t = 0$. It is subjected to the integral constraint.

$$\int_0^\infty x^2 \, dt = \tfrac{1}{4}.$$

Choose $u(t)$ so that

$$C = \int_0^\infty u^2 \, dt$$

is minimized. (Ans. $u = -e^{-2t}$.)

8. The state equations for a system are

$$\dot{x} = y, \qquad \dot{y} = u e^{\frac{1}{2}(1-u^2)}.$$

The control $u(t)$ is to be chosen so that the system is transferred from the state $x = y = 0$ to the state $x = 1$, $y = 0$ in the minimum time. Show that the motion separates into two phases during which u takes the constant values $+1$ and -1 respectively and that all necessary conditions, including the conditions (3.6.21) at the discontinuity, can be satisfied by taking the multipliers in the form

$$\lambda_x = -1, \qquad \lambda_y = t - 1.$$

Show that the minimal time is 2.

4 | HAMILTON-JACOBI EQUATION

4.1 Optimal cost function

Suppose that the problem formulated in section 3.1 has been solved for an initial state x^0 adopted at an instant t^0. Until now, x^0 and t^0 have been regarded as predetermined constants. We now change our attitude to these initial conditions and proceed to study the effect that changes in the values of x^0 and t^0 have on the subsequent optimal trajectory. In particular, the value of the optimal cost must depend upon the initial state and time, and we shall accordingly denote this *optimal cost function* by $V(x^0, t^0)$. Thus,

$$V(x^0, t^0) = \operatorname*{Min}_{u(t)} \left[G(x^1, t^1) + \int_{t^0}^{t^1} g(x, u, t)\, dt \right], \qquad (4.1.1)$$

where $u(t)$ ranges over all admissible controls which result in satisfaction of the constraints (3.1.1) (and any other constraints which we may later wish to impose upon the control vector).

x^0, t^0 together specify a point P_0 in a space S of $(M + 1)$ dimensions; this is the state space augmented by an additional dimension of time (*augmented state space*). The optimal behaviour of the system from the initial state x^0 will be given by an equation $x = x(t)$; this will determine a curve in S commencing at the point P_0 and terminating at a point P_1 on the intersection of the hypersurfaces whose equations are (3.1.1). The case when the state vector possesses a single component and there is only one constraint (3.1.1) is illustrated in Fig. 4.1. Clearly, any point in S can be called upon to play the role of the initial point P_0 and, provided the resulting optimization problem is uniquely soluble, a curve through the point representing the corresponding optimal behaviour will be defined. Further, any point (x', t') on such a curve can be taken as the initial point and, then, the part of the curve for which $t > t'$ will specify the system's optimal behaviour from this point. It follows that one curve, at most, will pass through each point of S and, hence, that a family of curves can be constructed in this way; such a family will be referred to as a *field of optimal trajectories*.

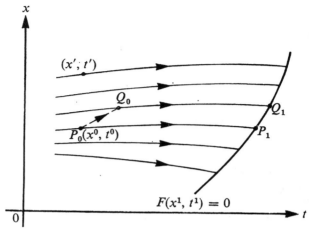

Fig. 4.1

4.2 Hamilton-Jacobi equation

Suppose the system is in an initial state x^0 at the instant t^0 and that, during the subsequent interval $(t^0, t^0 + \Delta t)$, the control vector is determined by an equation

$$u = v(t), \qquad (4.2.1)$$

where $v(t)$ is an arbitrary continuous function, satisfying all constraints imposed on the control vector but not necessarily leading to optimal behaviour. In the augmented state space S, the point representing the system's state moves from P_0 to Q_0 (Fig. 4.1) and the system's behaviour is specified by an equation

$$x = y(t). \qquad (4.2.2)$$

For $t > t^0 + \Delta t$, suppose $u(t)$ is optimal so that the behaviour of the system is represented by an optimal trajectory $Q_0 Q_1$. Then, the cost of this motion is given by

$$C = \int_{t^0}^{t^0 + \Delta t} g(y, v, t)\, dt + V(x^0 + \Delta x, t^0 + \Delta t), \qquad (4.2.3)$$

where $x^0 + \Delta x = y(t^0 + \Delta t)$. This cost will be minimal when $v(t)$ generates an optimal response and the point representing the system's state describes the trajectory $P_0 P_1$. Thus,

$$V(x^0, t^0) = \underset{v(t)}{\text{Min}} \left[\int_{t^0}^{t^0 + \Delta t} g(y, v, t) \, dt + V(x^0 + \Delta x, t^0 + \Delta t) \right]. \quad (4.2.4)$$

Now suppose that $\Delta t \to 0$ in the last equation. Then, it follows from the state equations that

$$\Delta x_i = f_i(x^0, v^0, t^0) \Delta t + O(\Delta t^2), \quad (4.2.5)$$

where $v^0 = v(t^0)$. Hence, employing Taylor's theorem, we calculate that

$$V(x^0 + \Delta x, t^0 + \Delta t)$$
$$= V[x^0 + f(x^0, v^0, t^0) \Delta t + O(\Delta t^2), t^0 + \Delta t]$$
$$= V(x^0, t^0) + \frac{\partial V}{\partial x_i} f_i(x^0, v^0, t^0) \Delta t + \frac{\partial V}{\partial t^0} \Delta t + O(\Delta t^2). \quad (4.2.6)$$

Also

$$\int_{t^0}^{t^0 + \Delta t} g(y, v, t) \, dt = g(x^0, v^0, t^0) \Delta t + O(\Delta t^2). \quad (4.2.7)$$

Substitution in equation (4.2.4) now leads to the result

$$\underset{v(t)}{\text{Min}} \left[g(x^0, v^0, t^0) + \frac{\partial V}{\partial x_i^0} f_i(x^0, v^0, t^0) + \frac{\partial V}{\partial t^0} + O(\Delta t) \right] = 0, \quad (4.2.8)$$

having cancelled $V(x^0, t^0)$ from both sides of the original equation and divided by Δt (N.B. $V(x^0, t^0)$ does not depend upon $v(t)$ and so is unaffected by the minimization operator). Letting $\Delta t \to 0$ in equation (4.2.8), it will be noted that, in the limit, the terms in the square bracket depend only upon v^0; further, $\partial V / \partial t^0$ is independent of v^0. Thus, the limiting form of the last equation is

$$\underset{v^0}{\text{Min}} \left[g(x^0, v^0, t^0) + \frac{\partial V}{\partial x_i^0} f_i(x^0, v^0, t^0) \right] = -\frac{\partial V}{\partial t^0}. \quad (4.2.9)$$

This is the *Hamilton-Jacobi equation*.

Since (x^0, t^0) can be taken to be any point in the field of optimal trajectories, it will usually be convenient to omit the zero superscripts. v will also be replaced by u and the Hamilton-Jacobi equation is therefore written in the form

$$\underset{u}{\text{Min}} \left[g(x, u, t) + \frac{\partial V}{\partial x_i} f_i(x, u, t) \right] = -\frac{\partial V}{\partial t}. \quad (4.2.10)$$

It is a partial differential equation of the first order satisfied by a variable V depending upon $(M + 1)$ independent variables x_1, $x_2, \ldots, x_M, t$. It is clear from the definition (4.1.1) of the optimal cost function that

$$V(x^1, t^1) = G(x^1, t^1), \tag{4.2.11}$$

where x^1, t^1 satisfy the end constraints (3.1.1); this provides a boundary condition upon the solution of the partial differential equation.

If an integral constraint of the type considered in section 3.5 is imposed upon the state and control vectors, this can be replaced by the differential constraint (3.5.2) in the manner explained in that section. The problem then reduces to one of the type we have been considering, the only effect being to augment the left-hand member of equation (4.2.10) by a term $\phi \partial V/\partial x_{M+1}$.

In the event that x, u are subject to a non-differential constraint of the type studied in section 3.4, the whole of the argument of the last two sections remains valid on the understanding that the minimization procedure is to be performed with respect to the set of control vectors u which satisfy this constraint at the point x, t of the augmented state space being considered.

It should be observed, therefore, that if we choose any point $P(x, t)$ of S, the optimal value of the control vector associated with the optimal trajectory through P can be determined by minimizing the expression

$$g(x, u, t) + \frac{\partial V}{\partial x_i} f_i(x, u, t) \tag{4.2.12}$$

with respect to the set of admissible values of the vector u satisfying all the non-differential constraints at P. This is a special case of *Pontryagin's Principle*, which will be further studied in section 4.4.

PROBLEM 26. Construct the Hamilton-Jacobi equation for the system described in problem 11 and, hence, determine its optimal behaviour.

Solution: In this case $f = u - 2x$, $g = x^2 + u^2/5$ and $G = 0$. The Hamilton-Jacobi equation is

$$\underset{u}{\text{Min}} \left[x^2 + \tfrac{1}{5}u^2 + \frac{\partial V}{\partial x}(u - 2x) \right] = -\frac{\partial V}{\partial t}.$$

The square bracket is a quadratic in u and is minimized by taking

$$u = -\frac{5}{2}\frac{\partial V}{\partial x}. \qquad (4.2.13)$$

Substituting for u, the equation becomes

$$\frac{5}{4}\left(\frac{\partial V}{\partial x}\right)^2 + 2x\frac{\partial V}{\partial x} - x^2 = \frac{\partial V}{\partial t}. \qquad (4.2.14)$$

It is required to find a solution of this equation satisfying the boundary condition (4.2.11) which, for this problem, takes the form

$$V(x^1, 1) = 0; \qquad (4.2.15)$$

since the final state is not subject to any constraint, this implies that V must vanish identically at $t = 1$. Such a boundary condition serves to determine the solution of equation (4.2.14) uniquely.

This unique solution can be found by assuming

$$V = x^2 T,$$

where T is a function of t alone. Substituting, equation (4.2.14) is found to be satisfied provided

$$\frac{dT}{dt} = 5T^2 + 4T - 1.$$

This equation is easily integrated to give

$$T = \frac{1 + Ae^{6t}}{5 - Ae^{6t}},$$

A being the constant of integration. The boundary condition (4.2.15) can be satisfied by requiring that $T = 0$ at $t = 1$; thus, $A = -e^{-6}$ and this gives

$$T = \frac{1 - e^{-6(1-t)}}{5 + e^{-6(1-t)}}.$$

Hence

$$V(x, t) = x^2 \frac{1 - e^{-6(1-t)}}{5 + e^{-6(1-t)}}$$

and the optimal control now follows from equation (4.2.13), namely

$$u = -5x\frac{1 - e^{-6(1-t)}}{5 + e^{-6(1-t)}}. \qquad (4.2.16)$$

Substituting this in the state equation, we obtain a first order differential equation for x, namely

$$\dot{x} = -3x \frac{5 - e^{-6(1-t)}}{5 + e^{-6(1-t)}}.$$

The integral of this equation satisfying the initial condition $x = x_0$ at $t = 0$ is

$$x = x_0 \frac{5e^{3(1-t)} + e^{-3(1-t)}}{5e^3 + e^{-3}}.$$

This agrees with the result already calculated in problem 11.

Substituting from this formula for x into equation (4.2.16), u can be calculated as a function of t. ●

4.3 Derivatives of the optimal cost function

In this section, it will be proved that

$$\frac{\partial V}{\partial x_i} = \lambda_i, \qquad \frac{\partial V}{\partial t} = -H, \tag{4.3.1}$$

where λ_i, H are the quantities defined along any optimal trajectory as explained in section 3.1.

For clarity of argument, it will be necessary to return to the notation in which the variables upon which V is dependent are denoted by x_i^0, t^0. Then,

$$V(x^0, t^0) = G(x^1, t^1) + \int_{t^0}^{t^1} g(x, u, t)\, dt, \tag{4.3.2}$$

where the vector functions $u(t)$, $x(t)$ correspond to the optimal trajectory originating at (x^0, t^0). These functions, together with the vector function $\lambda(t)$, satisfy Hamilton's equations (3.1.29) and the end conditions (3.1.30). x and u will also be dependent upon the initial point x^0, t^0, so we shall write

$$x = x(t, x^0, t^0), \qquad u = u(t, x^0, t^0). \tag{4.3.3}$$

Further, t^1 will depend upon this initial point, i.e.

$$t^1 = t^1(x^0, t^0). \tag{4.3.4}$$

At $t = t^1$, we know that $x = x^1$; hence,

$$x_j^1 = x_j[t^1(x^0, t^0), x^0, t^0] = x_j^1(x^0, t^0). \tag{4.3.5}$$

HAMILTON-JACOBI EQUATION

Also, at $t = t^0$, x must equal x^0; thus,

$$x_j^0 = x_j(t^0, x^0, t^0). \quad (4.3.6)$$

Differentiating (4.3.2) partially with respect to x_i^0, we find

$$\frac{\partial V}{\partial x_i^0} = \frac{\partial G}{\partial x_j^1}\frac{\partial x_j^1}{\partial x_i^0} + \frac{\partial G}{\partial t^1}\frac{\partial t^1}{\partial x_i^0} + g^1 \frac{\partial t^1}{\partial x_i^0}$$
$$+ \int_{t^0}^{t^1} \left(\frac{\partial g}{\partial x_j}\frac{\partial x_j}{\partial x_i^0} + \frac{\partial g}{\partial u_r}\frac{\partial u_r}{\partial x_i^0} \right) dt, \quad (4.3.7)$$

where $g^1 = g(x^1, u^1, t^1)$. Appealing to Hamilton's equations in the forms (3.6.6), (3.6.7), it will be seen that the integral in equation (4.3.7) can be expressed in the form

$$-\int_{t^0}^{t^1} \left[\left(\dot{\lambda}_j + \lambda_k \frac{\partial f_k}{\partial x_j} \right)\frac{\partial x_j}{\partial x_i^0} + \lambda_k \frac{\partial f_k}{\partial u_r}\frac{\partial u_r}{\partial x_i^0} \right] dt. \quad (4.3.8)$$

Since $x(t, x^0, t^0)$ must satisfy the system's state equations, it follows that

$$\frac{\partial x_k}{\partial t} = f_k(x, u, t) \quad (4.3.9)$$

will be an identity in the variables t, x^0 and t^0. Partially differentiating with respect to x_i^0, therefore, it follows that

$$\frac{\partial^2 x_k}{\partial x_i^0 \partial t} = \frac{\partial f_k}{\partial x_j}\frac{\partial x_j}{\partial x_i^0} + \frac{\partial f_k}{\partial u_r}\frac{\partial u_r}{\partial x_i^0}. \quad (4.3.10)$$

Thus, the integral (4.3.8) is equivalent to

$$-\int_{t^0}^{t^1} \frac{\partial}{\partial t}\left(\lambda_j \frac{\partial x_j}{\partial x_i^0} \right) dt = \left| \lambda_j \frac{\partial x_j}{\partial x_i^0} \right|_{t^1}^{t^0}. \quad (4.3.11)$$

Differentiating equation (4.3.6) partially with respect to x_i^0, we find that, at $t = t^0$, $\partial x_j/\partial x_i^0 = \delta_{ij}$ (Kronecker delta). Also, differentiating $x_j^1(x^0, t^0)$ as defined by equation (4.3.5), partially with respect to x_i^0, it follows that

$$\left(\frac{\partial x_j}{\partial t} \right)_{t=t^1} \frac{\partial t^1}{\partial x_i^0} + \left(\frac{\partial x_j}{\partial x_i^0} \right)_{t=t^1} = \frac{\partial x_j^1}{\partial x_i^0}. \quad (4.3.12)$$

110   ANALYTICAL METHODS OF OPTIMIZATION

Hence

$$\left|\lambda_j \frac{\partial x_j}{\partial x_i^0}\right|_{t^1}^{t^0} = \delta_{ij}\lambda_j^0 - \left[\frac{\partial x_j^1}{\partial x_i^0} - \left(\frac{\partial x_j}{\partial t}\right)_{t=t^1}\frac{\partial t^1}{\partial x_i^0}\right]\lambda_j^1$$

$$= \lambda_i^0 - \frac{\partial x_j^1}{\partial x_i^0}\frac{\partial J}{\partial x_j^1} + f_j^1 \lambda_j^1 \frac{\partial t^1}{\partial x_i^0}, \quad (4.3.13)$$

having employed equation (4.3.9) and the end conditions (3.1.30). Equation (4.3.7) can now be written

$$\frac{\partial V}{\partial x_i^0} = \lambda_i^0 + \frac{\partial}{\partial x_j^1}(G - J)\frac{\partial x_j^1}{\partial x_i^0} + \left(\frac{\partial G}{\partial t^1} + g^1 + f_j^1 \lambda_j^1\right)\frac{\partial t^1}{\partial x_i^0},$$

$$= \lambda_i^0 - \nu_m \frac{\partial F_m}{\partial x_j^1}\frac{\partial x_j^1}{\partial x_i^0} - F_m \frac{\partial \nu_m}{\partial x_j^1}\frac{\partial x_j^1}{\partial x_i^0} + \left(\frac{\partial G}{\partial t^1} + H^1\right)\frac{\partial t^1}{\partial x_i^0},$$

$$= \lambda_i^0 - \nu_m \frac{\partial F_m}{\partial x_j^1}\frac{\partial x_j^1}{\partial x_i^0} + \frac{\partial}{\partial t^1}(G - J)\frac{\partial t^1}{\partial x_i^0},$$

$$= \lambda_i^0 - \nu_m \left(\frac{\partial F_m}{\partial x_j^1}\frac{\partial x_j^1}{\partial x_i^0} + \frac{\partial F_m}{\partial t^1}\frac{\partial t^1}{\partial x_i^0}\right), \quad (4.3.14)$$

after referring to equations (3.1.1.), (3.1.21) and (3.1.30). Partial differentiation of the end constraints $F_m[x^1(x^0, t^0), t^1(x^0, t^0)] = 0$ with respect to x_i^0 shows that

$$\frac{\partial F_m}{\partial x_j^1}\frac{\partial x_j^1}{\partial x_i^0} + \frac{\partial F_m}{\partial t^1}\frac{\partial t^1}{\partial x_i^0} = 0 \quad (4.3.15)$$

and, hence, equation (4.3.14) finally yields the required result.

The second of the results (4.3.1) can be established by a similar argument. However, it follows immediately from the Hamilton-Jacobi equation (4.2.10). For, assuming that u is optimal, the left-hand member of the equation will be minimized and the equation shows that

$$\frac{\partial V}{\partial t} = -g - \frac{\partial V}{\partial x_i}f_i = -g - \lambda_i f_i = -H, \quad (4.3.16)$$

having used the result just established.

Only a slight amendment of the above argument need be made if the system is subject to additional constraints of the types studied in sections 3.4, 3.5. The results (4.3.1) continue to be valid in these circumstances.

4.4 Pontryagin's principle

A special case of this principle has been formulated at the end of section 4.2. It states that, at any point on an optimal trajectory, the optimal control vector can be found by minimizing the expression

$$P = g(x, u, t) + \frac{\partial V}{\partial x_i} f_i(x, u, t) \qquad (4.4.1)$$

with respect to the whole class of vectors u satisfying all non-differential constraints to which this vector is subject, x and t being regarded as fixed quantities (and hence $\partial V/\partial x_i$ are fixed). In the case when the performance index C is to be maximized, the expression P must, or course, also be maximized relative to u.

If u is only subject to the type of non-differential constraint considered in section 3.4, it follows from the results established in the previous section that $\partial V/\partial x_i = \lambda_i$ and, hence, that

$$P = g + \lambda_i f_i. \qquad (4.4.2)$$

If $\phi_\iota(x, u) = 0$ are the non-differential constraints, the Hamiltonian is

$$H = g + \lambda_i f_i + \mu_\iota \phi_\iota. \qquad (4.4.3)$$

Hence, for controls u satisfying the non-differential constraints, $P = H$ and Pontryagin's principle accordingly requires that the Hamiltonian be minimized (or maximized) with respect to all controls satisfying the non-differential constraints to which they are subject at any point (x, t) of the augmented control space. This principle can be employed in replacement of the weaker optimality conditions $\partial H/\partial u_j = 0$ (which the principle clearly implies).

However, the importance of Pontryagin's principle resides in the fact that it remains applicable when the control is subject to a wide variety of non-differential constraints. A rigorous proof of this statement will be found in the book of Pontryagin (*et al*) listed in the bibliography at the end of this text but, in the case of a constraint taking the form

$$\phi(u) \leqslant 0, \qquad (4.4.4)$$

a plausible argument in favour of the truth of the principle may be offered as follows:

Augment the integrand g of the cost C by a term $e^{\alpha\phi}$, where α is large and positive. If, now, u violates the constraint (4.4.4) over any subarc of a state trajectory, $e^{\alpha\phi}$ will take large values over this subarc and the value of C will be increased; the larger the value taken by α, the more pronounced will be this effect. It follows that, if we are attempting to minimize C, the constraint (4.4.4) will be satisfied automatically and may therefore be ignored. In fact, ϕ will always assume negative values, for if a certain control makes ϕ vanish, an arbitrarily small variation can be found which will make ϕ negative and this will result in a relatively large reduction in the value of C as $e^{\alpha\phi}$ changes from 1 to 0. However, if $\phi < 0$, $e^{\alpha\phi}$ will effectively vanish and will not contribute to C. Hence, the solution to the amended problem can be expected to approach the solution of the original problem as $\alpha \to +\infty$.

The Hamiltonian for the amended problem is

$$H' = g + \lambda_i f_i + e^{\alpha\phi}$$
$$= H + e^{\alpha\phi}, \qquad (4.4.5)$$

where H is the Hamiltonian for the original problem when the inequality is ignored. Since $\partial H'/\partial x_i = \partial H/\partial x_i$ and $\partial H'/\partial \lambda_i = \partial H/\partial \lambda_i$, Hamilton's equations may be constructed from H alone. Application of Pontryagin's principle to H' requires that H be minimized in such a way that $e^{\alpha\phi}$ is kept small. This will result in a control u for which $\phi < 0$. In the limit, as $\alpha \to +\infty$, this control u will minimize H under the constraint $\phi \leq 0$. The upshot is that, for our original problem, the optimal control u can be found by minimizing H subject to the constraint $\phi \leq 0$.

PROBLEM 27. (x, y) are state variables and u is the control variable for a system. The state equations are

$$\dot{x} = y + u, \qquad \dot{y} = -u,$$

and u is subject to the constraint $|u| \leq 1$. Show that the minimum time required to transfer the system from the state $x = y = 1$ to the state $x = y = 0$ is $\sqrt{10} + 1$.

Solution: For this problem $g = 1$ and the Hamiltonian is

$$H = 1 + \lambda_x(y + u) - \lambda_y u.$$

The equations for the multipliers are therefore,

$$\dot\lambda_x = 0, \quad \dot\lambda_y = -\lambda_x,$$

and these have general solutions

$$\lambda_x = a, \quad \lambda_y = b - at,$$

where a, b are constants.

According to Pontryagin's principle, H must be minimized relative to controls u satisfying $|u| \leqslant 1$. Since $H = (\lambda_x - \lambda_y)u +$ terms not involving u, there are three cases to consider. (i) If $(\lambda_x - \lambda_y) > 0$, then $u = -1$. (ii) If $\lambda_x - \lambda_y = 0$, then u is indeterminate. (iii) If $(\lambda_x - \lambda_y) < 0$, then $u = +1$. But, $\lambda_x - \lambda_y = a - b + at$, which is linear in t. It follows that $(\lambda_x - \lambda_y)$ can change sign at most once during the optimal motion and, therefore, that there can be at most one "switching point" at which u changes from $+1$ to -1 or vice versa.

Let us investigate the trajectories in state space relating to the control $u = +1$. Along these arcs, $\dot x = y + 1$ and $\dot y = -1$; thus $dx/dy = -y - 1$ and

$$x = \text{constant} - \tfrac{1}{2}y^2 - y.$$

This is the equation of a family of parabolas having a common axis $y = -1$.

Similarly, we can show that the state space trajectories corresponding to the control $u = -1$ form another family of parabolas with equation

$$x = \text{constant} + \tfrac{1}{2}y^2 - y;$$

these curves have a common axis $y = 1$.

One curve from each family passes through the initial point $(1, 1)$ and these are indicated in Fig. 4.2. The two curves through the final point $(0, 0)$ are also shown in the figure. Arrows show the directions in which these arcs are described as t increases (the senses of the arrows are determined by the equations $\dot y = -1$ on $u = +1$ and $\dot y = +1$ on $u = -1$); only those parts of the curves which leave $(1, 1)$ and arrive at $(0, 0)$ are relevant to our problem and have been drawn.

Since there can be at most one "junction point" where a switch

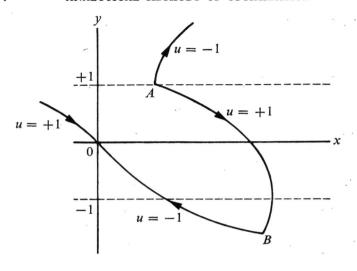

Fig. 4.2

is made from a curve of one family to a curve of the other, this point must be B. It is the intersection of the parabolas

$$x = \tfrac{5}{2} - \tfrac{1}{2}y^2 - y, \qquad x = \tfrac{1}{2}y^2 - y$$

and is easily found to have coordinates $x = \tfrac{5}{4} + \sqrt{\tfrac{5}{2}},\ y = -\sqrt{\tfrac{5}{2}}$.

Over AB, $\dot{y} = -1$ and y decreases by $1 + \sqrt{(5/2)}$. Hence, the time to describe this arc $1 + \sqrt{(5/2)}$. Over BO, $\dot{y} = +1$ and y increases by $\sqrt{(5/2)}$. Thus, the time for this arc is $\sqrt{(5/2)}$. The total time for the optimal trajectory is accordingly $\sqrt{10} + 1$.

$(\lambda_x - \lambda_y)$ must vanish at the junction B. Also, since the transit time is variable, $H = 0$ at the final point. These two conditions serve to fix the constants a and b. It is left for the reader to verify that $a = 2/\sqrt{10},\ b = 1 + 4/\sqrt{10}$.

PROBLEM 28. In the oscillatory system described in problem 15, the mass is to be transferred from its initial state $x = 0$, $\dot{x} = 1$ to the final equilibrium state $x = \dot{x} = 0$ in minimum time. The control u is subject to the constraint $|u| \leqslant 1$. Prove that the minimum time is $\tfrac{1}{4}\pi + \cos^{-1}(1/2\sqrt{2})$.

Solution: The canonical state equations are $\dot{x} = y$, $\dot{y} = u - x$ and the Hamiltonian is accordingly

$$H = 1 + \lambda_x y + \lambda_y(u - x).$$

Hamilton's equations are

$$\dot{\lambda}_x = \lambda_y, \qquad \dot{\lambda}_y = -\lambda_x;$$

the general solution for λ_y is $\lambda_y = a \sin(t + \alpha)$.

Applying Pontryagin's principle to H, it follows that $u = +1$ whenever λ_y is negative and $u = -1$ whenever λ_y is positive. Since λ_y changes sign at instants separated by intervals which are all of duration π, if there is more than one switching point, it follows that the period of transit between the initial and final states will exceed π. If, therefore, there exists an optimal trajectory with one switching point, for which the transit time is less than π, no other trajectories need be considered (we are assuming that a trajectory without switching points cannot be found to satisfy the end conditions).

If $u = +1$, then $\dot{x} = y$, $\dot{y} = 1 - x$ and, solving, it follows that

$$x = 1 - b \cos(t + \beta), \qquad y = b \sin(t + \beta).$$

These are parametric equations of a circle in state space having radius b and centre $(1, 0)$; the circle is described with unit angular velocity in a clockwise sense as t increases. Similarly, the family of trajectories associated with the control $u = -1$ is a set of concentric circles with centre at $(-1, 0)$, also described in a clockwise sense.

Assuming there is only one switching point, an optimal trajectory through the initial point A and final point O can be constructed from a circular arc AB with centre at C $(-1, 0)$ and a circular arc BO with centre at $D(1, 0)$ (Fig. 4.3). This will be shown to correspond to a transit time of less than π and, hence, is the optimal trajectory required (it is easily verified that no other trajectory with a single switching point exists, for which the transit time is less than π).

In the triangle BCD, $BC = AC = \sqrt{2}$, $BD = OD = 1$, $CD = 2$. The cosine rule then gives: angle $CBD = \cos^{-1}(-1/2\sqrt{2})$. Since the angle ACX is $\pi/4$, this is the time required for the system to move from the state A to the state X. If the arcs XB, BO subtend angles θ, ϕ at their centres, the time for the system to move from X to O

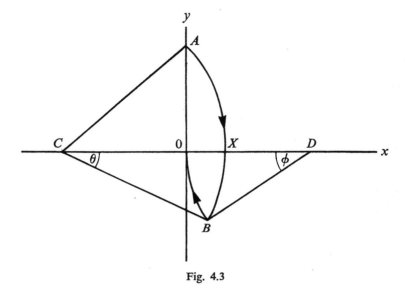

Fig. 4.3

is $(\theta + \phi) = \pi -$ angle $CBD = \cos^{-1}(1/2\sqrt{2})$. Thus, the total time of transit from A to O is $\tfrac{1}{4}\pi + \cos^{-1}(1/2\sqrt{2})$. ●

4.5 Optimal rocket trajectories

An important application of Pontryagin's principle is to the calculation of rocket trajectories of minimal propellent expenditure in a given gravitational field (air resistance will be neglected so that the resulting theory will only be applicable to rockets moving in interplanetary space).

Denoting by M the mass of the rocket at time t, the thrust T generated by the motor is given by

$$T = -c\frac{dM}{dt}, \qquad (4.5.1)$$

where c is the exhaust velocity.

Let $Ox_1x_2x_3$ be any convenient rectangular cartesian inertial frame and let $g_i(x_1, x_2, x_3, t)$ ($i = 1, 2, 3$) be the components of gravitational acceleration at a point x_i at time t (the dependence upon t will be due to the motions of the various attracting bodies belonging to the solar system). Then, if l_i are the components of a unit vector in the direction of the thrust T and v_i are the components of rocket velocity, the rocket's equations of motion can be written

$$\dot{v}_i = \frac{cm}{M} l_i + g_i, \tag{4.5.2}$$

$$\dot{x}_i = v_i, \tag{4.5.3}$$

$$\dot{M} = -m, \tag{4.5.4}$$

m denoting the rate of expenditure of propellent.

Equations (4.5.2)–(4.5.4) are the state equations for the system being studied, v_i, x_i, M being the seven state variables and l_i, m being the four control variables. The control variables are subject to a non-differential constraint, viz.

$$l_i l_i = l_1^2 + l_2^2 + l_3^2 = 1. \tag{4.5.5}$$

It will also be assumed that m cannot exceed the value m_0 and, hence, that it is subject to the inequality constraints

$$0 \leqslant m \leqslant m_0. \tag{4.5.6}$$

The problem to be solved is that of transferring the rocket between two given terminals A, B, having coordinates x_i^0, x_i^1, respectively, the rocket's velocity at these points to take predetermined values v_i^0, v_i^1, respectively and the mass of the rocket to have a known value M^0 at the departure terminal A; the cost to be minimized is the overall propellent expenditure, i.e. the final mass M' is to be maximized. Thus,

$$C = G = M^1. \tag{4.5.7}$$

The times of departure and arrival, t^0, t^1, will be regarded as predetermined.

We first construct the Hamiltonian in the form

$$H = \lambda_i \left(\frac{cm}{M} l_i + g_i \right) + \mu_i v_i - \nu m, \tag{4.5.8}$$

where λ_i, μ_i, ν are multipliers satisfying the Hamilton equations

$$\dot{\lambda}_i = -\frac{\partial H}{\partial v_i} = -\mu_i, \tag{4.5.9}$$

$$\dot{\mu}_i = -\frac{\partial H}{\partial x_i} = -\lambda_j \frac{\partial g_j}{\partial x_i}, \tag{4.5.10}$$

$$\dot{\nu} = -\frac{\partial H}{\partial M} = \frac{cm}{M^2} \lambda_i l_i. \tag{4.5.11}$$

Since M^1 is open to choice, the corresponding multiplier satisfies the end condition

$$v^1 = \frac{\partial G}{\partial M^1} = 1; \qquad (4.5.12)$$

the remaining multipliers are subject to no end conditions.

According to Pontryagin's principle, H has to be maximized relative to the controls l_i and m, subject to the constraints (4.5.5), (4.5.6). This implies that κm has to be maximized, where

$$\kappa = \frac{c}{M} \lambda_i l_i - v. \qquad (4.5.13)$$

Considering first the variable m, there are three cases. (i) If $\kappa > 0$, then $m = m_0$; (ii) if $\kappa < 0$, then $m = 0$; (iii) if $\kappa = 0$, m is not determined. Any arc along which $m = m_0$ will be termed an MT (maximum thrust) arc. An arc along which $m = 0$ will be called an NT (null thrust) arc. Usually, the optimal trajectory will comprise arcs of these two types only, κ being the *thrust magnitude switching function* whose change of sign signals a change from an arc of one type to an arc of the other. It is possible, in certain circumstances, however, for κ to vanish identically along an arc; m can then take values intermediate between the limits 0 and m_0; such arcs are called IT (intermediate thrust) arcs (see Problem 36, p. 146).

Turning now to the maximization of H with respect to the l_i, it is necessary to maximize κ (equation (4.5.13)) with respect to these variables, under the constraint (4.5.5). Since $\lambda_i l_i$ is the scalar product of two vectors with components λ_i, l_i, the first vector being regarded as given and the second vector being variable but having a fixed unit magnitude, it is obvious that $\lambda_i l_i$ is maximized by aligning the vector l_i with the vector λ_i. Thus, we take

$$\lambda_i = p l_i, \qquad (4.5.14)$$

where p is the magnitude of the vector λ_i. The vector λ_i is called the *primer* and its direction determines the direction of the optimal rocket thrust on MT and IT arcs. Substituting for λ_i from equation (4.5.14) into equation (4.5.13), we obtain

$$\kappa = \frac{cp}{M} - v \qquad (4.5.15)$$

as the final expression for the switching function.

The equations determining the components of the primer can be found by elimination of the μ_i between equations (4.5.9), (4.5.10); the result is

$$\ddot{\lambda}_i = \lambda_j \frac{\partial g_j}{\partial x_i}. \tag{4.5.16}$$

Denoting the primer vector by **p** and the gravitational acceleration by **g**, this equation can also be expressed in the form

$$\ddot{\mathbf{p}} = \nabla(\mathbf{p} \cdot \mathbf{g}). \tag{4.5.17}$$

Differentiation of equation (4.5.15) and use of equations (4.5.11), (4.5.14), gives the result

$$\dot{\kappa} = c\dot{p}/M. \tag{4.5.18}$$

On an NT arc, M remains constant and the last equation can be integrated to yield

$$\kappa = \frac{cp}{M} + \text{constant}, \tag{4.5.19}$$

revealing that ν is constant along an arc of this type. On an IT arc, $\kappa = 0$ and it follows that p is constant along such an arc.

In the case of a space rocket driven by a conventional motor employing chemical propellents, m_0 is so large that the phases of maximum thrust are comparatively short (of the order of a few seconds). The movement of the rocket in the interplanetary gravitational field during such a phase can then be neglected and the motor thrust be regarded as impulsive. Since κ is positive over an MT arc and negative over an NT arc, the form of its graph against t during a phase of maximum thrust is as indicated in Fig. 4.4 (a) (we shall suppose IT arcs absent from the optimal trajectory throughout this discussion). In circumstances where the duration of the MT phase is negligible, the graph must take the limiting from shown in Fig. 4.4(b), i.e. $\kappa = \dot{\kappa} = 0$ at the instant of the impulsive thrust.

Now consider the NT arc joining a pair of consecutive points of impulsive thrust (junction points). Along this arc, equation (4.5.19) is valid and, since κ vanishes at both junctions, p must assume the same value P at these points. It now follows that $p = P$ at all junction points on the optimal trajectory. Also, since $\nu = 1$ (equation (4.5.12)) at the arrival terminal and ν and M are constant along the NT arc (if any) leading to this terminal, we deduce that, at the last junction

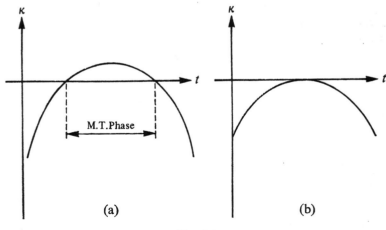

Fig. 4.4

point, $\kappa = cP/M^1 - 1$. Hence

$$P = M^1/c. \tag{4.5.20}$$

Equation (4.5.18) shows that $\dot{p} = 0$ at a junction point, implying that M^1/c is the maximum value of p.

The necessary conditions for an optimal trajectory can now be expressed entirely in terms of the primer vector **p** as follows:
  (a) **p** satisfies equation (4.5.17),
  (b) whenever the thrust is operative it must be aligned with **p**,
  (c) the magnitude of **p** assumes a maximum value M^1/c at all junction points (i.e. $\dot{p} = 0$),
  (d) the primer and its first time derivative are continuous everywhere.

Further properties of optimal rocket trajectories are established in the book by Lawden listed in the bibliography at the end of this book.

4.6 Weierstrass and Clebsch conditions

In the case when the control vector is not subject to inequality constraints, Pontryagin's principle is equivalent to a necessary condition due to Weierstrass. It follows from Pontryagin's principle that if $u(t)$ is the optimal control at time t, $x(t)$ is the associated optimal state and $\lambda(t)$ is the adjoint multiplier vector defined along the optimal trajectory, then

$$H(x, v, \lambda, t) \geqslant H(x, u, \lambda, t) \qquad (4.6.1)$$

for all non-optimal controls v satisfying the non-differential constraints. The inequality is reversed if C is to be maximized. Since H is given by equation (4.4.3) and $\phi_i(x, u, t) = \phi_i(x, v, t) = 0$, the inequality (4.6.1) is equivalent to

$$g(x, v, t) - g(x, u, t) + \lambda_i[f_i(x, v, t) - f_i(x, u, t)] \geqslant 0. \qquad (4.6.2)$$

This is *Weierstrass's condition*.

Another necessary condition which can be derived from Pontryagin's principle is that due to Clebsch. We shall assume that the control is subject to neither inequality nor equality constraints. Then, since $H(x, v, \lambda, t)$ is minimized when $v = u$, it follows from results established in section 1.2 that $\partial H/\partial v$ vanishes (as already known) and that the quadratic form

$$\frac{\partial^2 H}{\partial v_r \, \partial v_s} \xi_r \xi_s \qquad (4.6.3)$$

is positive semi-definite when $v = u$. Thus, the matrix with elements $\partial^2 H/\partial u_r \partial u_s$ must be positive semi-definite. This matrix will be denoted by H_{uu}. If C is to be maximized, then H_{uu} must be negative semi-definite. This is the *Clebsch condition*. Evidently, this condition is a consequence of the Pontryagin principle but, unlike the Weierstrass condition, is not equivalent to it (being less stringent).

Exercises 4

1. A point moving along the x-axis has its acceleration controlled to be u, where u is subject to the constraint $|u| \leqslant 1$. It is desired to move the point from the state $x = \dot{x} = 0$ to the state $x = 1$, $\dot{x} = 0$ in the minimum time. Employ Pontryagin's principle to determine the motion and hence show that the minimum time is 2

2. In problem 28, replace the stated initial state by $x = 1$, $\dot{x} = 0$ and rework the calculation Show that the minimum time is now $\pi - \cos^{-1}(\tfrac{1}{4})$.

3. A second order system has state equations

$$\dot{x} = y, \qquad \dot{y} = x + u,$$

and the control variable u is subject to the inequality constraint $|u| \leqslant 1$. The system is to be transferred from the state (x_0, y_0) to the state (x_1, y_1) in the minimum time. Using Pontryagin's principle, show that u takes one of its extreme values throughout the motion and that the value of u cannot change more than once. Show that the optimal trajectories in the xy-plane are the families of rectangular hyperbolae

$$(x + 1)^2 - y^2 = \text{constant}, \qquad (x - 1)^2 - y^2 = \text{constant}.$$

If $(x_0, y_0) = (-\frac{1}{2}, 0)$, $(x_1, y_1) = (\frac{1}{2}, 0)$, prove that the minimal time of transfer is $2 \cosh^{-1} 2 = \log(7 + 4\sqrt{3})$.

4. The acceleration α of a point P which moves along the x-axis can be controlled between the limits $\alpha = -1$ and $\alpha = +1$. At time $t = 0$, P is at the point $x = 1$ with velocity $\dot{x} = 1$. At time $t = 4$, it is to be brought to the origin with zero velocity and the motion is to be such that

$$C = \int_0^4 |\alpha|\, dt$$

is a minimum. Prove that the optimal motion has three phases, (i) $\alpha = -1$, (ii) $\alpha = 0$, (iii) $\alpha = +1$, which occur in this order and sketch the corresponding families of trajectories in the xy-plane ($y = \dot{x}$). Calculate the times of the two switches separating the phases and show that the minimum value of C is $4 - \sqrt{3}$.

5. A system is governed by the second order equation

$$\ddot{x} + \dot{x} = u,$$

u being the control quantity. u is subject to the constraint $|u| \leq 1$. The system is to be transferred from the state $x = 1$, $\dot{x} = 0$ to the state $x = \dot{x} = 0$ in the minimum time. Prove that the optimal motion separates into two phases (i) $u = -1$ and (ii) $u = +1$, and that the minimal time is $2 \log(\sqrt{e} + \sqrt{(e-1)})$.

6. The equations of rectilinear motion of a rocket operating at constant power can be put into the form $\dot{x} = u$, $\dot{y} = u^2$, where x is the velocity, y is the reciprocal of the mass and u is the acceleration due to the motor thrust. u is subject to the constraint $|u| \leq 1$ and is to be chosen so as to minimize the time of transit between given states (x_0, y_0), (x_1, y_1), where $y_1 > y_0$. Prove that u either takes the constant value $(y_1 - y_0)/(x_1 - x_0)$ or alternates between its extreme values $+1$ and -1 ("bang-bang" control). In the former case, show that the transit time is $(x_1 - x_0)^2/(y_1 - y_0)$, and in the latter case, show that this time is $(y_1 - y_0)$.

5 | THE ACCESSORY OPTIMIZATION PROBLEM

5.1 Second variation of the cost

In the special case of static systems studied in Chapter 1, in addition to the first necessary conditions (1.2.6), for the cost C to be minimized there is a second set of necessary conditions arising from the requirement that the second derivative of the cost with respect to the parameter ϵ must be positive. In this chapter, a necessary condition for the optimization of a dynamic system issuing from the same source will be studied.

It will be assumed that the optimization problem has been posed in the form and using the notation given in section 3.1. It will also be assumed that the optimal control vector $u(t)$ is continuous for all relevant values of t and, therefore, that the optimal state vector $x(t)$ possesses a continuous first derivative.

We first postulate an embedding theorem in the following form: There exist families of control vectors $v(t, \epsilon)$ and state vectors $y(t, \epsilon)$, depending upon a single parameter ϵ, which satisfy the state equations for all values of ϵ in some neighbourhood Δ of $\epsilon = 0$ and which reduce to the optimal control vector $u(t)$ and state vector $x(t)$ for $\epsilon = 0$. y satisfies the end conditions

$$y(t^0, \epsilon) = x^0, \qquad y(t^1, \epsilon) = x^1, \qquad (5.1.1)$$

where x^1, t^1 are taken to be fixed at their optimal values, for all values of ϵ in Δ. All partial derivatives of $v(t, \epsilon), y(t, \epsilon)$, up to and including those of the third order, are continuous functions of t and ϵ over a sufficiently extensive region of $t\epsilon$-space to validate our subsequent calculations, except that $\partial v/\partial \epsilon$ may be discontinuous (finitely) with respect to t for a finite number of values of this variable in the interval (t^0, t^1) (this will induce corresponding discontinuities with respect to t in $\partial^2 y/\partial t \partial \epsilon$, $\partial^3 y/\partial t (\partial \epsilon)^2$, via the state equations).

Given any value of ϵ in Δ, the cost C of the corresponding system behaviour will be determined and, if C is minimal for $\epsilon = 0$, the following conditions must be satisfied at $\epsilon = 0$,

$$\frac{dC}{d\epsilon} = 0, \qquad \frac{d^2C}{d\epsilon^2} \geqslant 0 \qquad (5.1.2)$$

(the existence of the derivatives can be deduced from the assumptions made above). The values of the derivatives $dC/d\epsilon$, $d^2C/d\epsilon^2$ at $\epsilon = 0$ are termed the *first* and *second variations* of the cost respectively, with respect to the family of variations defined earlier.

First and second variations of the control and state vectors are now defined by the equations

$$\xi(t) = \left(\frac{\partial v}{\partial \epsilon}\right)_{\epsilon=0}, \qquad \eta(t) = \left(\frac{\partial y}{\partial \epsilon}\right)_{\epsilon=0}, \qquad (5.1.3)$$

$$\omega(t) = \left(\frac{\partial^2 v}{\partial \epsilon^2}\right)_{\epsilon=0}, \qquad \zeta(t) = \left(\frac{\partial^2 y}{\partial \epsilon^2}\right)_{\epsilon=0}, \qquad (5.1.4)$$

respectively. Since v, y satisfy the state equations for all values of ϵ in the neighbourhood of zero, the equation

$$\frac{\partial y_i}{\partial t} = f_i(y, v, t) \qquad (5.1.5)$$

may be partially differentiated with respect to ϵ, and ϵ put equal to zero, to yield equations of the first variation

$$\dot{\eta}_i = \frac{\partial f_i}{\partial x_j}\eta_j + \frac{\partial f_i}{\partial u_r}\xi_r, \qquad (5.1.6)$$

valid along the optimal trajectories $u(t)$, $x(t)$. A second partial differentiation with respect to ϵ and setting ϵ to zero, leads to the equations of second variation

$$\dot{\zeta}_i = \frac{\partial^2 f_i}{\partial x_j \partial x_k}\eta_j\eta_k + 2\frac{\partial^2 f_i}{\partial x_j \partial u_r}\eta_j\xi_r + \frac{\partial^2 f_i}{\partial u_r \partial u_s}\xi_r\xi_s + \frac{\partial f_i}{\partial x_j}\zeta_j + \frac{\partial f_i}{\partial u_r}\omega_r. \qquad (5.1.7)$$

Since ξ may be discontinuous for a finite number of values of t, both $\dot{\eta}$ and $\dot{\zeta}$ will be discontinuous for such values of t. However, η and ζ are both continuous.

Partial differentiation of the end conditions (5.1.1) leads to the following end conditions for the variations of the state vector:

$$\eta(t^0) = \eta^0 = 0, \qquad \eta(t^1) = \eta^1 = 0, \tag{5.1.8}$$

$$\zeta(t^0) = \zeta^0 = 0, \qquad \zeta(t^1) = \zeta^1 = 0. \tag{5.1.9}$$

The cost expressed as a function of ϵ is given by

$$C(\epsilon) = G(y^1, t^1) + \int_{t^0}^{t^1} g[y(t, \epsilon), v(t, \epsilon), t] \, dt. \tag{5.1.10}$$

Since y^1, t^1 are fixed under variation of ϵ, G is independent of ϵ and a first differentiation with respect to this parameter gives

$$\frac{dC}{d\epsilon} = \int_{t^0}^{t^1} \left(\frac{\partial g}{\partial y_j} \frac{\partial y_j}{\partial \epsilon} + \frac{\partial g}{\partial v_r} \frac{\partial v_r}{\partial \epsilon} \right) dt. \tag{5.1.11}$$

A second differentiation then yields

$$\frac{d^2C}{d\epsilon^2} = \int_{t^0}^{t^1} \left(\frac{\partial^2 g}{\partial y_j \partial y_k} \frac{\partial y_j}{\partial \epsilon} \frac{\partial y_k}{\partial \epsilon} + 2 \frac{\partial^2 g}{\partial y_j \partial v_r} \frac{\partial y_j}{\partial \epsilon} \frac{\partial v_r}{\partial \epsilon} \right.$$
$$\left. + \frac{\partial^2 g}{\partial v_r \partial v_s} \frac{\partial v_r}{\partial \epsilon} \frac{\partial v_s}{\partial \epsilon} + \frac{\partial g}{\partial y_j} \frac{\partial^2 y_j}{\partial \epsilon^2} + \frac{\partial g}{\partial v_r} \frac{\partial^2 v_r}{\partial \epsilon^2} \right) dt. \tag{5.1.12}$$

Putting $\epsilon = 0$, we find that

$$\left(\frac{d^2C}{d\epsilon^2} \right)_0 = \int_{t^0}^{t^1} \left(\frac{\partial^2 g}{\partial x_j \partial x_k} \eta_j \eta_k + 2 \frac{\partial^2 g}{\partial x_j \partial u_r} \eta_j \xi_r \right.$$
$$\left. + \frac{\partial^2 g}{\partial u_r \partial u_s} \xi_r \xi_s + \frac{\partial g}{\partial x_j} \zeta_j + \frac{\partial g}{\partial v_r} \omega_r \right) dt. \tag{5.1.13}$$

Multiplying equation (5.1.7) through by λ_i, integrating over the interval (t^0, t^1) and adding the result to the last equation, we obtain the equation

$$\left(\frac{d^2C}{d\epsilon^2} \right)_0 + \int_{t^0}^{t^1} \lambda_i \dot{\zeta}_i \, dt$$
$$= \int_{t^0}^{t^1} \left(\frac{\partial^2 H}{\partial x_j \partial x_k} \eta_j \eta_k + 2 \frac{\partial^2 H}{\partial x_j \partial u_r} \eta_j \xi_r + \frac{\partial^2 H}{\partial u_r \partial u_s} \xi_r \xi_s \right.$$
$$\left. + \frac{\partial H}{\partial x_j} \zeta_j + \frac{\partial H}{\partial u_r} \omega_r \right) dt, \tag{5.1.14}$$

where
$$H = g + \lambda_i f_i \tag{5.1.15}$$
is the Hamiltonian.

Along the optimal trajectory, Hamilton's equations
$$\dot{\lambda}_i = -\frac{\partial H}{\partial x_i}, \quad \frac{\partial H}{\partial u_r} = 0, \tag{5.1.16}$$
are known to be valid. It follows that equation (5.1.14) reduces to the form
$$\left(\frac{d^2 C}{d\epsilon^2}\right)_0 = \int_{t^0}^{t^1} \left(2A - \frac{d}{dt}(\lambda_i \zeta_i)\right) dt, \tag{5.1.17}$$
where
$$2A = \frac{\partial^2 H}{\partial x_j \partial x_k} \eta_j \eta_k + 2 \frac{\partial^2 H}{\partial x_j \partial u_r} \eta_j \xi_r + \frac{\partial^2 H}{\partial u_r \partial u_s} \xi_r \xi_s. \tag{5.1.18}$$
But, λ_i and ζ_i are continuous functions of t and it accordingly follows that
$$\left(\frac{d^2 C}{d\epsilon^2}\right)_0 = \int_{t^0}^{t^1} 2A \, dt, \tag{5.1.19}$$
having integrated the derivative $d(\lambda_i \zeta_i)/dt$ and used the end conditions (5.1.9).

Equation (5.1.19) is the required formula for the second variation of the cost.

5.2 Accessory problem. Conjugate points

Making use of the form for the second variation of C calculated in the last section, it follows that a necessary condition for minimization of C is
$$\int_{t^0}^{t^1} A \, dt \geqslant 0. \tag{5.2.1}$$

The variations $\xi_r(t)$ are arbitrary except that the associated variations $\eta_i(t)$ determined by equations (5.1.6) must satisfy the end conditions (5.1.8); further, the $\xi_r(t)$ are permitted to have only a finite number of finite discontinuities.

An *accessory minimization problem* now suggests itself in the following form: Regard $\xi(t)$ as the control vector, $\eta(t)$ as the state vector and the equations (5.1.6) as state equations; the accessory problem is to minimize the cost given by

$$\mathscr{C} = \int_{t^0}^{t^1} A\, dt, \tag{5.2.2}$$

subject to the end constraints (5.1.8). Assuming that $u(t)$ defines an optimal control for the original problem, it follows from the condition (5.2.1) that the minimal value of the cost for the accessory problem is zero (this value can certainly be attained by taking $\xi = 0$, $\eta = 0$).

The Hamiltonian for the accessory problem is clearly given by

$$\mathscr{H} = A + \Lambda_i\left(\frac{\partial f_i}{\partial x_j}\eta_j + \frac{\partial f_i}{\partial u_r}\xi_r\right), \tag{5.2.3}$$

where Λ_i are the multipliers for the problem. This leads to Hamilton equations of the form

$$\dot{\Lambda}_i = -\frac{\partial \mathscr{H}}{\partial \eta_i} = -\frac{\partial^2 H}{\partial x_i\, \partial x_j}\eta_j - \frac{\partial^2 H}{\partial x_i\, \partial u_r}\xi_r - \Lambda_j\frac{\partial f_j}{\partial x_i}, \tag{5.2.4}$$

$$\frac{\partial \mathscr{H}}{\partial \xi_r} = \frac{\partial^2 H}{\partial x_i\, \partial u_r}\eta_i + \frac{\partial^2 H}{\partial u_r\, \partial u_s}\xi_s + \Lambda_i\frac{\partial f_i}{\partial u_r} = 0. \tag{5.2.5}$$

These equations, together with the equations of variation (5.1.6), are linear in the variables η_i, ξ_r, Λ_i, and determine these quantities uniquely provided the values of η_i, Λ_i are known for a single value of t (it is here necessary to assume that the $N \times N$ determinant with elements $\partial^2 H/\partial u_r \partial u_s$ does not vanish identically over the range of values of t being considered). Let us suppose that this system of equations (called the *accessory equations*) possesses a solution $\eta(t)$ over the interval (t^0, t') ($t^0 < t' < t^1$), satisfying the end conditions

$$\eta_i(t^0) = \eta_i(t') = 0, \tag{5.2.6}$$

and not vanishing identically over the interval. If such a solution can be found, the point on the state-space trajectory $x_i = x_i(t)$ where $t = t'$ is said to be *conjugate* to the initial point where $t = t^0$. Defining $\eta_i(t)$, $\xi_r(t)$, $\Lambda_i(t)$ over the interval (t^0, t') by this solution and over the interval (t', t^1) by the equations

$$\eta_i(t) = 0, \quad \xi_r(t) = 0, \quad \Lambda_i(t) = 0, \tag{5.2.7}$$

we have constructed an admissible variation of the trajectories $x = x(t)$, $u = u(t)$ ($\eta(t)$ is continuous throughout the interval (t^0, t^1) and $\xi(t)$ may have a discontinuity at $t = t'$, but is continuous elsewhere). For this variation, it follows from equations (5.1.6), (5.1.18), (5.2.4), (5.2.5), that

$$2A = -\eta_i \dot{\Lambda}_i - \dot{\eta}_i \Lambda_i, \qquad (5.2.8)$$

over the interval (t^0, t'). Hence, since η_i, Λ_i are continuous over this interval.

$$\mathscr{C} = -\int_{t^0}^{t'} \frac{d}{dt}(\eta_i \Lambda_i)\, dt = |\eta_i \Lambda_i|_{t^0}^{t'}. \qquad (5.2.9)$$

But η vanishes at both limits, showing that $\mathscr{C} = 0$. It follows that this variation minimizes $\mathscr{C}$ and therefore must satisfy all the necessary conditions for a minimum in the accessory problem. One of these conditions is that, at a discontinuity of the control $\xi(t)$, $\Lambda(t)$ must be continuous (section 3.6); hence $\Lambda(t') = 0$. Thus, η and Λ take known values (viz. zero) at $t = t'$ and, therefore, $\eta(t)$, $\Lambda(t)$ are uniquely determined throughout the interval (t^0, t') as solutions of the equations (5.1.6), (5.2.4) and (5.2.5). But these equations evidently possess the solution $\eta(t) = 0$, $\Lambda(t) = 0$, $\xi(t) = 0$, satisfying the conditions $\eta = 0$, $\Lambda = 0$ at $t = t'$. We conclude that $\eta(t)$ must vanish identically over the interval (t^0, t') and so have arrived at a contradiction.

If, therefore, the functions $x = x(t)$, $u = u(t)$, minimize C, it is a necessary condition that there should be no point conjugate to the initial point $t = t^0$ on the trajectory $x = x(t)$ within the interval (t^0, t^1) (*Jacobi's condition*).

In the special case of a linear system with quadratic cost function (equation (2.7.1)), the Hamiltonian is given by equation (2.7.2) and it will be found that the accessory equations are identical in form with the Hamilton equations (2.7.3)–(2.7.5), the variables η_i, ξ_r, Λ_i taking the places of x_i, u_r, λ_i respectively. We conclude that, for such a system, the Jacobi necessary condition is satisfied provided it is impossible to drive the system from an initial state $x^0 = 0$ at $t = t^0$ to a final state $x' = 0$ at $t = t'$, where $t^0 < t' \leqslant t^1$, by application of a control $u(t)$ satisfying the Hamilton equations.

PROBLEM 29. Show that the solution to problem 16 satisfies the necessary condition formulated above.

Solution: The Hamilton equations for Problem 16 have been shown to be

$$\dot{x} = y, \qquad \dot{y} = -x + u,$$
$$\dot{\lambda}_x = \lambda_y, \qquad \dot{\lambda}_y = -\lambda_x,$$
$$\lambda_y + 2u = 0.$$

These equations are all linear and the general solutions for x and y are found to be

$$x = (\alpha t + \beta)\cos t + (\gamma t + \delta)\sin t,$$
$$y = (\gamma t + \alpha + \delta)\cos t - (\alpha t + \beta - \gamma)\sin t.$$

These are also the general solutions of the accessory equations for η_x, η_y, respectively. Choosing the parameters α, β, γ, δ so that $x = y = 0$ at the initial instant $t = 0$, it is found that

$$x = \alpha t \cos t + (\gamma t - \alpha)\sin t,$$
$$y = \gamma t \cos t - (\alpha t - \gamma)\sin t.$$

The condition that x, y also vanish for some other value of t in $(0, \tfrac{1}{2}\pi)$ is

$$\begin{vmatrix} t\cos t - \sin t & t\sin t \\ -t\sin t & t\cos t + \sin t \end{vmatrix} = 0$$

for this value of t. This condition reduces to $t^2 - \sin^2 t = 0$; since this equation possesses only one real root $t = 0$, we have shown that there is no point on the solution trajectory between $t = 0$ and $t = \tfrac{1}{2}\pi$ which is conjugate to the initial point. Jacobi's necessary condition for an optimal trajectory is accordingly satisfied. ●

5.3 Solution of the accessory equations

The Hamilton equations for the original problem can be written

$$\dot{x}_i = \frac{\partial H}{\partial \lambda_i} = f_i, \tag{5.3.1}$$

$$\dot{\lambda}_i = -\frac{\partial H}{\partial x_i}, \tag{5.3.2}$$

$$0 = \frac{\partial H}{\partial u_r}. \tag{5.3.3}$$

The general solution of this system of equations will involve $2M$ constants of integration. Let α be a typical such constant. Then, variation of α will yield a family of solutions which can be written

$$x = x(t, \alpha), \quad \lambda = \lambda(t, \alpha), \quad u = u(t, \alpha). \quad (5.3.4)$$

Substituting from these equations into the equations (5.3.1)–(5.3.3), the Hamilton equations will be satisfied identically in both t and α. Hence, differentiating equations (5.3.1)–(5.3.3) partially with respect to α, we obtain the equations

$$\frac{\partial}{\partial t}\left(\frac{\partial x_i}{\partial \alpha}\right) = \frac{\partial f_i}{\partial x_j}\frac{\partial x_j}{\partial \alpha} + \frac{\partial f_i}{\partial u_r}\frac{\partial u_r}{\partial \alpha}, \quad (5.3.5)$$

$$\frac{\partial}{\partial t}\left(\frac{\partial \lambda_i}{\partial \alpha}\right) = -\frac{\partial^2 H}{\partial x_i \partial x_j}\frac{\partial x_j}{\partial \alpha} - \frac{\partial^2 H}{\partial x_i \partial u_r}\frac{\partial u_r}{\partial \alpha} - \frac{\partial^2 H}{\partial x_i \partial \lambda_j}\frac{\partial \lambda_j}{\partial \alpha}, \quad (5.3.6)$$

$$0 = \frac{\partial^2 H}{\partial x_i \partial u_r}\frac{\partial x_i}{\partial \alpha} + \frac{\partial^2 H}{\partial u_r \partial u_s}\frac{\partial u_s}{\partial \alpha} + \frac{\partial^2 H}{\partial u_r \partial \lambda_i}\frac{\partial \lambda_i}{\partial \alpha}. \quad (5.3.7)$$

Noting that

$$\frac{\partial^2 H}{\partial x_i \partial \lambda_j} = \frac{\partial f}{\partial x_i}, \quad \frac{\partial^2 H}{\partial u_r \partial \lambda_i} = \frac{\partial f_i}{\partial u_r}, \quad (5.3.8)$$

it will be observed that equations (5.3.5)–(5.3.7) have the same form as the accessory equations (5.1.6), (5.2.4), (5.2.5). It follows that

$$\eta_i = \frac{\partial x_i}{\partial \alpha}, \quad \Lambda_i = \frac{\partial \lambda_i}{\partial \alpha}, \quad \xi_r = \frac{\partial u_r}{\partial \alpha}, \quad (5.3.9)$$

is a solution of the accessory equations.

By this means, $2M$ independent solutions of the accessory equations can be constructed along an optimal trajectory determined by given values of the parameters α. Since the system of accessory equations is linear and of order $2M$, their general solution along the optimal trajectory can then be written down by superposition.

PROBLEM 30. A system has state variable x and control variable θ and its state equation is $\dot{x} = \tan \theta$. It is to be transferred from the initial state $x = x_0$ at $t = 0$ to the final state $x = x_1$ at $t = T$ in such a way that

$$C = \int_0^T x \sec \theta \, dt$$

is minimized.

Solution: The Hamiltonian is given by

$$H = x \sec \theta + \lambda \tan \theta$$

and Hamilton's equations are

$$\dot{x} = \tan \theta, \qquad \dot{\lambda} = -\sec \theta,$$
$$x \sec \theta \tan \theta + \lambda \sec^2 \theta = 0, \quad \text{or} \quad \lambda = -x \sin \theta.$$

Elimination of λ, yields the equation $x\dot\theta = 1$. Whence,

$$\frac{1}{x}\frac{dx}{d\theta} = \frac{\dot{x}}{x\dot\theta} = \tan \theta.$$

Integrating, we find that $x = \alpha \sec \theta$, where α is a constant of integration. It now follows that $\dot\theta \sec \theta = 1/\alpha$ and a further integration gives

$$\log(\sec \theta + \tan \theta) = (t + \beta)/\alpha,$$

where β is a second integration constant. Thus

$$\tan \theta = \sinh \frac{t + \beta}{\alpha}, \qquad \sec \theta = \cosh \frac{t + \beta}{\alpha},$$

from which we deduce that

$$x = \alpha \cosh \frac{t + \beta}{\alpha}, \qquad \lambda = -\alpha \sinh \frac{t + \beta}{\alpha}. \qquad (5.3.13)$$

The constants α, β can now be calculated by use of the end conditions. Consider the family of trajectories in the xt-plane satisfying the initial condition $x = x_0$ at $t = 0$. This family is sketched in Fig. 5.1 for the case $x_0 > 0$; it possesses an envelope E as shown. It is clear that, unless the final point (T, x_1) lies in the region above (or on) E, no trajectory belonging to the family satisfies the end conditions. If the final point lies above E, there will be two members of the family satisfying these end conditions. If the final point lies on E, only one member of the family can be found to satisfy the end conditions.

Assuming that values of α, β can be found to satisfy the end conditions, by partial differentiation of the first of equations (5.3.13), it is

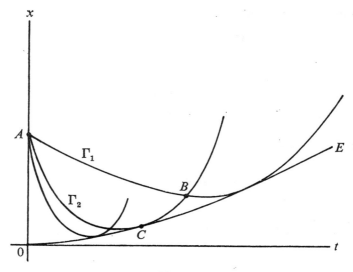

Fig. 5.1

found that

$$\frac{\partial x}{\partial \alpha} = \cosh \tau - \tau \sinh \tau,$$

$$\frac{\partial x}{\partial \beta} = \sinh \tau,$$

where $\tau = (t + \beta)/\alpha$. The general solution of the accessory equations for the variation η of x can now be written down, namely,

$$\eta = A(\cosh \tau - \tau \sinh \tau) + B \sinh \tau. \tag{5.3.14}$$

Suppose η vanishes for $t = 0$ and for some later value of t. The condition for this to be possible can be expressed in the form

$$\frac{\beta}{\alpha} - \coth \frac{\beta}{\alpha} = \frac{t + \beta}{\alpha} - \coth \frac{t + \beta}{\alpha}. \tag{5.3.15}$$

This is the condition that the two points $t = 0$, t on the trajectory (α, β) should be conjugate.

Now, for trajectories through the initial point $A(0, x_0)$, α and β are related by the equation

$$\alpha \cosh(\beta/\alpha) = x_0. \tag{5.3.16}$$

Differentiating with respect to α, we find that

$$\frac{d\beta}{d\alpha} = \frac{\beta}{\alpha} - \coth\frac{\beta}{\alpha}. \tag{5.3.17}$$

To calculate the equation of the envelope E, it is necessary to differentiate the first of equations (5.3.13) partially with respect to α (β being treated as a known function of α). The result may be written

$$\frac{d\beta}{d\alpha} = \frac{t+\beta}{\alpha} - \coth\frac{t+\beta}{\alpha}; \tag{5.3.18}$$

the equations (5.3.13), (5.3.16)–(5.3.18) then determine E. In particular, eliminating $d\beta/d\alpha$ between equations (5.3.17), (5.3.18), we arrive at equation (5.3.15) and this equation can accordingly be interpreted as fixing the value of t where the trajectory (α, β) touches the envelope E. We conclude that this point of contact is the point on the trajectory (α, β) conjugate to the initial point A (e.g. the point C in Fig. 5.1 is conjugate to A).

Suppose $B(T, x_1)$ represents the final point. There are two possible trajectories Γ_1, Γ_2 passing through A and B. It is clear from the figure that one of these trajectories Γ_1 does not have contact with E at a point on its arc AB, whereas the other trajectory Γ_2 touches E at a point C on its arc AB. Thus, Γ_2 does not satisfy the necessary condition for an optimal trajectory and only Γ_1 can be optimal. ●

5.4 The accessory Riccati equation

The accessory equations (5.1.6), (5.2.4), (5.2.5), will first be expressed in matrix form. Put $\xi = [\xi_1, \xi_2, \ldots, \xi_N]^T$, $\eta = [\eta_1, \eta_2, \ldots, \eta_M]^T$, $\Lambda = [\Lambda_1, \Lambda_2, \ldots, \Lambda_M]^T$ and denote the $M \times M$ matrices with ij-elements $\partial f_i/\partial x_j$, $\partial^2 H/\partial x_i \partial x_j$ by f_x, H_{xx}, the $M \times N$ matrices with ir-elements $\partial f_i/\partial u_r$, $\partial^2 H/\partial x_i \partial u_r$ by f_u, H_{xu} and the $N \times N$ matrix with rs-element $\partial^2 H/\partial u_r \partial u_s$ by H_{uu}. Then the accessory equations can be written

$$\dot{\eta} = f_x \eta + f_u \xi, \tag{5.4.1}$$

$$\dot{\Lambda} = -H_{xx}\eta - H_{xu}\xi - f_x^T \Lambda, \tag{5.4.2}$$

$$H_{xu}^T \eta + H_{uu}\xi + f_u^T \Lambda = 0. \tag{5.4.3}$$

From this point, it will be assumed that the matrix H_{uu} is positive definite at every point on the extremal trajectory being considered from $t = t^0$ to $t = t^1$ (any trajectory $x = x(t)$, $u = u(t)$, $\lambda = \lambda(t)$, which satisfies the Hamilton equations but is not necessarily optimal, will be referred to as an *extremal*). Thus, the determinant $|H_{uu}|$ cannot vanish at any point of the interval $[t^0, t^1]$ and the inverse matrix H_{uu}^{-1} accordingly exists at all points. Solving equation (5.4.3) for ξ and substituting in the other two equations, we now obtain

$$\dot{\eta} = (f_x - f_u H_{uu}^{-1} H_{xu}^T)\eta - f_u H_{uu}^{-1} f_u^T \Lambda, \tag{5.4.4}$$

$$\dot{\Lambda} = (H_{xu} H_{uu}^{-1} H_{xu}^T - H_{xx})\eta + (H_{xu} H_{uu}^{-1} f_u^T - f_x^T)\Lambda. \tag{5.4.5}$$

This is a system of $2M$ first order linear differential equations in the $2M$ variables η_i, Λ_i.

Let $\eta^{(i)}(t)$, $\Lambda^{(i)}(t)$ ($i = 1, 2, \ldots, M$) be a set of M solutions of these equations. Using these columns, we construct square matrices $N(t)$, $L(t)$ of order M, thus

$$N = [\eta^{(1)} \mid \eta^{(2)} \mid \cdots \mid \eta^{(M)}], \qquad L = [\Lambda^{(1)} \mid \Lambda^{(2)} \mid \cdots \mid \Lambda^{(M)}]. \tag{5.4.6}$$

Then, N and L will also satisfy the accessory equations to give

$$\dot{N} = (f_x - f_u H_{uu}^{-1} H_{xu}^T)N - f_u H_{uu}^{-1} f_u^T L, \tag{5.4.7}$$

$$\dot{L} = (H_{xu} H_{uu}^{-1} H_{xu}^T - H_{xx})N + (H_{xu} H_{uu}^{-1} f_u^T - f_x^T)L. \tag{5.4.8}$$

At points where $|N| \neq 0$, a matrix K, also of order M, can now be defined by the equation

$$K = LN^{-1}. \tag{5.4.9}$$

To find the differential equation satisfied by K, we put $L = KN$ in equation (5.4.8) and employ equation (5.4.7) to eliminate $\dot{N}$; the result is,

$$[\dot{K} - (Kf_u + H_{xu})H_{uu}^{-1}(f_u^T K + H_{xu}^T) + Kf_x + f_x^T K + H_{xx}]N = 0. \tag{5.4.10}$$

Multiplying on the right by N^{-1}, we find that K satisfies the Riccati equation

$$\dot{K} - (Kf_u + H_{xu})H_{uu}^{-1}(f_u^T K + H_{xu}^T) + Kf_x + f_x^T K + H_{xx} = 0. \tag{5.4.11}$$

This is the *Accessory Riccati Equation*.

In particular, let us study the set of solutions $\eta_j^{(i)}(t)$, $\Lambda_j^{(i)}(t)$, of the accessory equations generated by the initial conditions

$$\eta_j^{(i)}(t^0) = 0, \qquad \Lambda_j^{(i)}(t^0) = \delta_j{}^i, \tag{5.4.12}$$

where $\delta_j{}^i$ are the Kronecker deltas. Let $\eta_i(t)$, $\Lambda_i(t)$ be any other solution satisfying the initial conditions $\eta_i = 0$ at $t = t^0$. Defining η_i^*, Λ_i^* by the equations

$$\begin{aligned}\eta_i^*(t) &= \eta_i(t) - \eta_i^{(j)}(t)\Lambda_j(t^0), \\ \Lambda_i^*(t) &= \Lambda_i(t) - \Lambda_i^{(j)}(t)\Lambda_j(t^0),\end{aligned} \tag{5.4.13}$$

it follows that η_i^*, Λ_i^* provide a solution of the accessory equations such that $\eta_i^*(t^0) = 0$, $\Lambda_i^*(t^0) = 0$. But this must be the null solution and, therefore,

$$\eta_i(t) = \eta_i^{(j)}(t)\Lambda_j^0, \qquad \Lambda_i(t) = \Lambda_i^{(j)}(t)\Lambda_j^0. \tag{5.4.14}$$

Now suppose that $\eta_i(t') = 0$ ($i = 1, 2, \ldots, M$), where $t' > t^0$. Then, either $\Lambda_j^0 = 0$ ($j = 1, 2, \ldots, M$) and η, Λ is the null solution, or $|\eta_j^{(i)}(t')| = 0$. We conclude that a point conjugate to t^0 exists in the interval $[t^0, t^1]$ if, and only if,

$$|\eta_j^{(i)}| = 0 \tag{5.4.15}$$

at the conjugate point.

Suppose that no point conjugate to the initial point t^0 exists in the closed interval $[t^0, t^1]$. It may be proved that the relationship between a pair of conjugate points is a continuous one, in the sense that if one of the pair is shifted along its extremal by a small displacement then so is the other. If, therefore, the first point conjugate to t^0 to be encountered as we move along the extremal in the sense of t increasing lies at t', where $t' > t^1$, it will be possible to find a positive ϵ such that no point conjugate to $t^0 - \epsilon$ lies in the interval $t^0 - \epsilon < t \leqslant t^1$. Hence, redefining the functions $\eta_j^{(i)}(t)$, $\Lambda_j^{(i)}(t)$ to satisfy the conditions (5.4.12) at $t = t^0 - \epsilon$ instead of at $t = t^0$, we can prove that $|\eta_j^{(i)}|$ does not vanish at any point of the closed interval $[t^0, t^1]$. Employing this set of solutions of the accessory equations to construct the matrix N, the inverse N^{-1} will exist throughout the interval $[t^0, t^1]$. It then follows that the accessory Riccati equation (5.4.11) must possess a solution matrix K (given by equation (5.4.9)) which exists (i.e. is finite) at all points of the interval $[t^0, t^1]$. Further, this matrix K will be symmetric, as we shall now prove.

Let $P = K^{-1}$; then, by equation (5.4.9), $P = NL^{-1}$ and, hence, $N = PL$. Thus, substituting for N in equation (5.4.7) and eliminating $\dot{L}$ by use of equation (5.4.8), it may be proved that P satisfies the Riccati equation

$$\dot{P} + (PH_{xu} + f_u)H_{uu}^{-1}(H_{xu}^T P + f_u^T) \\ - PH_{xx}P - Pf_x^T - f_x P = 0. \quad (5.4.16)$$

By transposition of this equation, we find that P^T satisfies the same Riccati equation. Hence, if $P = P^T$ for some value of t, then this equation will be valid for all values of t (inspection of the Riccati equation (5.4.11) reveals that K possesses the same property). Consider the matrix P constructed from the set of solutions of the accessory equations determined by the initial conditions (5.4.12) satisfied at $t = t^0 - \epsilon$. At the initial point, $N = 0$ and $L = I$; hence, at this point $P = 0$ and $P = P^T$ is true. We conclude that P is symmetric for all values of t and, hence, that its inverse K is also symmetric.

To summarize, if no point conjugate to the initial point t^0 exists in the interval $[t^0, t^1]$, then the accessory Riccati equation has a finite symmetric solution throughout this interval. Equivalently, if the accessory Riccati equation possesses no such solution over the interval $[t^0, t^1]$, then a point conjugate to t^0 must exist in this interval and the extremal under consideration cannot be optimal.

PROBLEM 31. Write down and solve the accessory Riccati equation for Problem 30.

Solution: The Hamiltonian for the problem is $H = x \sec \theta + \lambda \tan \theta$, x being the state variable and θ the control variable. Thus,

$$H_{xx} = 0, \quad H_{xu} = \sec \theta \tan \theta,$$
$$H_{uu} = x \sec \theta (2 \tan^2 \theta + 1) + 2\lambda \sec^2 \theta \tan \theta.$$

The state equation is $\dot{x} = \tan \theta$, from which it follows that $f_x = 0$, $f_u = \sec^2 \theta$.

Along a nominal optimal trajectory

$$\tan \theta = \sinh \frac{t + \beta}{\alpha}, \quad \sec \theta = \cosh \frac{t + \beta}{\alpha},$$
$$x = \alpha \cosh \frac{t + \beta}{\alpha}, \quad \lambda = -\alpha \sinh \frac{t + \beta}{\alpha},$$

where α, β are constants determined by the end constraints. Hence,

$$H_{xx} = 0, \qquad H_{xu} = \sinh\frac{t+\beta}{\alpha}\cosh\frac{t+\beta}{\alpha},$$

$$H_{uu} = \alpha\cosh^2\frac{t+\beta}{\alpha}, \qquad f_x = 0, \qquad f_u = \cosh^2\frac{t+\beta}{\alpha}.$$

The accessory Riccati equation is therefore

$$\frac{dK}{d\tau} = (K\cosh\tau + \sinh\tau)^2,$$

where $\tau = (t+\beta)/\alpha$.

A particular solution of the Riccati equation is easily verified to be $K = -\coth\tau$. The general solution can then be found by changing the dependent variable to y by putting $K = -\coth\tau + 1/y$. We find that

$$K = \frac{(\tau+\gamma)\cosh\tau - \sinh\tau}{\cosh\tau - (\tau+\gamma)\sinh\tau},$$

where γ is the constant of integration.

It will be seen that K becomes infinite for values of τ such that

$$\coth\tau - \tau = \gamma.$$

This equation always has two roots in τ (one positive and one negative) for any value of γ. These roots always define a pair of conjugate points on the trajectory (α, β) for, if a solution $\eta(\tau)$ of the accessory equations can be found which vanishes at $\tau = \tau_1$ and $\tau = \tau_2$, it follows from equation (5.3.14) that

$$A(\cosh\tau_1 - \tau_1\sinh\tau_1) + B\sinh\tau_1 = 0,$$
$$A(\cosh\tau_2 - \tau_2\sinh\tau_2) + B\sinh\tau_2 = 0,$$

and the conditions for these equations to have a non-null solution in A, B is that

$$\coth\tau_1 - \tau_1 = \coth\tau_2 - \tau_2.$$

Thus, the condition that an extremal arc does not contain a pair of conjugate points is equivalent to the condition that the Riccati equation possesses a solution which is finite over the arc.

5.5 Sufficiency conditions

If we can prove that $(d^2C/d\epsilon^2)_0$ is strictly positive for all admissible variations $\xi(t)$, $\eta(t)$ satisfying the equations of variation (5.1.6) and the end conditions (5.1.8), it will follow that the control $u(t)$ yields at least a local minimum of C with respect to variations which do not affect the end states x^0, x^1. It will now be shown that conditions which are sufficient for this purpose are: (i) H_{uu} positive definite at every point on the extremal (the *Clebsch condition*), (ii) no point on the extremal is conjugate to the initial point t^0 (the *Jacobi condition*).

Employing the notation of the previous section, we can write equation (5.1.19) in the form

$$\left(\frac{d^2C}{d\epsilon^2}\right)_0 = \int_{t^0}^{t^1} (\xi^T H_{uu} \xi + 2\eta^T H_{xu} \xi + \eta^T H_{xx} \eta) \, dt, \quad (5.5.1)$$

where ξ, η is any admissible variation satisfying the equations of variation (5.4.1) and the end conditions $\eta^0 = \eta^1 = 0$. Satisfaction of the Jacobi condition implies that the accessory Riccati equation possesses a symmetric solution $K(t)$ which exists (i.e. remains finite) over the whole of the closed interval $[t^0, t^1]$. Employing this solution, we deduce from equation (5.4.1) that

$$2\eta^T K(\dot{\eta} - f_x \eta - f_u \xi) = 0. \quad (5.5.2)$$

This equation is equivalent to the equation

$$\frac{d}{dt}(\eta^T K \eta) = \eta^T \dot{K} \eta + 2\eta^T K f_x \eta + 2\eta^T K f_u \xi. \quad (5.5.3)$$

Integrating over the interval $[t^0, t^1]$ and using the end conditions $\eta^0 = \eta^1 = 0$, this leads to the result

$$\int_{t^0}^{t^1} (\eta^T \dot{K} \eta + 2\eta^T K f_x \eta + 2\eta^T K f_u \xi) \, dt = 0. \quad (5.5.4)$$

Adding equations (5.5.1) and (5.5.4), we find that

$$\left(\frac{d^2C}{d\epsilon^2}\right)_0 = \int_{t^0}^{t^1} [\xi^T H_{uu} \xi + 2\eta^T (H_{xu} + K f_u) \xi + \eta^T (\dot{K} + 2K f_x + H_{xx}) \eta] \, dt. \quad (5.5.5)$$

But, by transposition it follows that

$$\eta^T (H_{xu} + K f_u) \xi = \xi^T (H_{xu}{}^T + f_u{}^T K) \eta, \quad (5.5.6)$$

$$\eta^T K f_x \eta = \eta^T f_x{}^T K \eta. \quad (5.5.7)$$

Thus equation (5.5.5) can be written in the form

$$\left(\frac{d^2C}{d\epsilon^2}\right)_0 = \int_{t^0}^{t^1} [\xi^T H_{uu}\xi + \eta^T(H_{xu} + Kf_u)\xi + \xi^T(H_{xu}{}^T + f_u{}^T K)\eta$$
$$+ \eta^T(\dot{K} + Kf_x + f_x{}^T K + H_{xx})\eta] \, dt. \quad (5.5.8)$$

Since K satisfies equation (5.4.11), the last equation reduces to

$$\left(\frac{d^2C}{d\epsilon^2}\right)_0 = \int_{t^0}^{t^1} [\xi^T H_{uu}\xi + \eta^T(H_{xu} + Kf_u)\xi + \xi^T(H_{xu}{}^T + f_u{}^T K)\eta$$
$$+ \eta^T(Kf_u + H_{xu})H_{uu}{}^{-1}(f_u{}^T K + H_{xu}{}^T)\eta] \, dt$$
$$= \int_{t^0}^{t^1} \theta^T H_{uu} \theta \, dt. \quad (5.5.9)$$

where

$$\theta = \xi + H_{uu}{}^{-1}(f_u{}^T K + H_{xu}{}^T)\eta. \quad (5.5.10)$$

Since, by the Clebsch condition, H_{uu} is positive definite, equation (5.5.9) shows that $(d^2C/d\epsilon^2)_0$ can only vanish for variations which make θ identically zero; for all other variations, $(d^2C/d\epsilon^2)_0 > 0$. But $\theta = 0$ implies that

$$\xi = -H_{uu}{}^{-1}(f_u{}^T K + H_{xu}{}^T)\eta \quad (5.5.11)$$

and this equation, together with the equation of variation (5.4.1) and the accessory Riccati equation (5.4.11), imply that ξ, η and $\Lambda = K\eta$ satisfy the accessory equations (5.4.1)–(5.4.3); this may be proved thus: Multiplying the Riccati equation on the right by η and using equation (5.5.11), we first obtain the equation

$$\dot{K}\eta + (Kf_u + H_{xu})\xi + (Kf_x + f_x{}^T K + H_{xx})\eta = 0. \quad (5.5.12)$$

By equation (5.4.1), this last equation reduces to

$$\frac{d}{dt}(K\eta) + H_{xu}\xi + H_{xx}\eta + f_x{}^T K\eta = 0 \quad (5.5.13)$$

and then, putting $\Lambda = K\eta$, equation (5.4.2) emerges. Hence, any variation which satisfies equation (5.5.11), also provides a solution of the accessory equations. But this solution would satisfy the end conditions $\eta^0 = \eta^1 = 0$ and its existence would accordingly imply that the end points were conjugate, contrary to our original hypothesis that the Jacobi condition is satisfied. We conclude that equation (5.5.11) cannot be valid over the interval $[t^0, t^1]$ and, hence,

140   ANALYTICAL METHODS OF OPTIMIZATION

that $(d^2C/d\epsilon^2)_0$ is strictly positive for all admissible variations. Thus, C_0 is a local minimum for all admissible variations.

PROBLEM 32. Show that the length of a great circle path between two points on the surface of a sphere satisfies the sufficiency conditions for a local minimum, provided the arc is not greater than, or equal to, a semi-circle.

Solution: Let θ, ϕ be latitude and longitude respectively of a point on the sphere and choose the equator so that the end points are $O(\theta = \phi = 0)$ and $A(\theta = 0, \phi = \alpha)$. Let the path joining O, A be specified by an equation of the form $\theta = F(\phi)$; this is equivalent to the differential equation

$$\frac{d\theta}{d\phi} = u(\phi), \qquad (5.5.14)$$

where $u = dF/d\phi$, together with the end conditions $\theta = 0$ at $\phi = 0, \alpha$.

Taking the radius of the sphere as the unit of length, the length of the path OA is C, where

$$C = \int_0^\alpha \sqrt{(u^2 + \cos^2 \theta)}\, d\phi. \qquad (5.5.15)$$

We now regard u as the control variable, θ as the state variable and let ϕ play the role of the time. C is to be minimized. The Hamiltonian is therefore

$$H = \sqrt{(u^2 + \cos^2 \theta)} + \lambda u \qquad (5.5.16)$$

and Hamilton's equations are accordingly

$$\frac{d\lambda}{d\phi} = -\frac{\partial H}{\partial \theta} = \frac{\sin \theta \cos \theta}{\sqrt{(u^2 + \cos^2 \theta)}}, \qquad (5.5.17)$$

$$0 = \frac{\partial H}{\partial u} = \frac{u}{\sqrt{(u^2 + \cos^2 \theta)}} + \lambda. \qquad (5.5.18)$$

Eliminating u and $d\phi$ between equations (5.5.14), (5.5.17) and (5.5.18), we get

$$\frac{\lambda}{1 - \lambda^2} d\lambda = -\tan \theta\, d\theta.$$

Integrating, this yields

$$\lambda = -\sqrt{(1 - P \sec^2 \theta)},$$

where P is a constant of integration (we assume u positive and λ negative). u can now be calculated from equation (5.5.18) and it then follows from equations (5.5.14) that

$$\frac{d\phi}{d\theta} = \frac{\sec^2\theta}{\sqrt{(\beta^2 - \tan^2\theta)}},$$

where $\beta^2 = (1 - P)/P$. A second integration leads to the equation of the family of geodesies, namely,

$$\tan\theta = \beta\sin(\phi + \gamma), \qquad (5.5.19)$$

γ being the constant of integration. Evidently $\beta = 0$ for the geodesic through the points O and A, i.e. it is the equator $\theta = 0$.

Along the great circle $\theta = 0$, we now calculate that $\lambda = u = 0$. Hence,

$$\left.\begin{array}{ll} H_{\theta\theta} = -1, \quad H_{\theta u} = 0, \quad H_{uu} = 1, \\ f_\theta = 0, \quad f_u = 1. \end{array}\right\} \qquad (5.5.20)$$

Clearly, the Clebsch condition is satisfied. The accessory equations along the great circle are

$$d\eta/d\phi = \xi, \quad d\Lambda/d\phi = \eta, \quad \xi + \Lambda = 0. \qquad (5.5.21)$$

Thus, $d^2\eta/d\phi^2 + \eta = 0$ and, if η is to vanish at the initial point $\phi = 0$, it follows that $\eta = Q\sin\phi$. At a point conjugate to the initial point, η must vanish; hence, $\phi = \pi$ and the conjugate point lies at the opposite pole to O. We conclude that the Jacobi condition is satisfied provided the arc OA is less than a semi-circle and that, in this case, OA will provide at least a local minimum for the length of the path connecting O and A. If OA is a semi-circle, so that O, A are opposite poles of the sphere, there are any number of great circle paths between O and A, all of the same length, and no one of these can provide a local minimum for the path length. If OA is greater than a semi-circle, the minor great circle arc between O and A provides the shortest path.

The accessory Riccati equation for the arc OA takes the form

$$\frac{dK}{d\phi} = K^2 + 1.$$

Its solution is evidently $K = \tan(\phi + \epsilon)$. For no choice of ϵ can this solution be finite over the whole of the arc OA if $\alpha > \pi$. This confirms that the Jacobi condition is not satisfied for such an arc. ●

PROBLEM 33. A linear system has state equations

$$\dot{x} = y - x, \quad \dot{y} = u - y,$$

u being the control. The system is to be taken from the state $x = y = 0$ at $t = 0$ to a final state $x = y = 1$ at $t = 1$, with minimization of the index

$$C = \int_0^1 u^2 \, dt.$$

Obtain the optimal control and show that the sufficiency conditions for a local minimum of C are satisfied.

Solution: The Hamiltonian is

$$H = u^2 + \lambda(y - x) + \mu(u - y)$$

and equations governing the optimal control are

$$\dot{\lambda} = \lambda, \quad \dot{\mu} = \mu - \lambda, \quad 2u + \mu = 0.$$

Solving these equations for u, we find that $u = (A + Bt)e^t$, A and B being constants of integration.

Extremal arcs in the state space have therefore to satisfy the equations

$$\dot{x} = y - x, \quad \dot{y} = -y + (A + Bt)e^t.$$

By integrating these equations, it is found that the family of extremals through the initial point $x = y = 0$ at $t = 0$, is given by

$$\begin{aligned} x &= [\alpha + \beta(t-1)]e^t - [\alpha(2t+1) - (t+1)\beta]e^{-t}, \\ y &= [2\alpha + \beta(2t-1)]e^t - (2\alpha - \beta)e^{-t}, \end{aligned} \quad (5.5.22)$$

where we have put $A = 4\alpha$, $B = 4\beta$. This family contains a member such that $x = y = 0$ at time $t (\neq 0)$ and for which α, β are not both zero, if and only if

$$\begin{vmatrix} e^t - (2t+1)e^{-t} & (t-1)e^t + (t+1)e^{-t} \\ 2e^t - 2e^{-t} & (2t-1)e^t + e^{-t} \end{vmatrix} = 0.$$

This condition reduces to

$$\cosh 2t = 2t^2 + 1.$$

But, for non-vanishing t, the power series expansion of $\cosh 2t$ reveals that $\cosh 2t > 2t^2 + 1$. It follows that x and y cannot both vanish at $t = 0$ and at some later instant. Thus, the Jacobi condition is satisfied for any extremal arc.

The matrix H_{uu} has but one element, namely 2. The Clebsch condition is obviously satisfied. We have therefore verified that the sufficiency conditions for a local minimum of C are both satisfied.

To satisfy the end conditions $x = y = 1$ at $t = 1$, it follows from equations (5.5.22) that

$$\alpha = \frac{e(e^2 - 1)}{e^4 - 6e^2 + 1}, \quad \beta = -\frac{e(e^2 + 1)}{e^4 - 6e^2 + 1}.$$

Hence, the optimal control is given by

$$u = \frac{4e[(1 - t)e^2 - 1 - t]}{e^4 - 6e^2 + 1} e^t$$

and the minimum cost is then found to be

$$C_{\min} = \frac{4e^2(e^2 - 1)}{e^4 - 6e^2 + 1}. \quad \bullet$$

5.6 Singular arcs

One of the sufficiency conditions we have required to be satisfied, if a given extremal is to minimize the cost C, is the Clebsch condition that H_{uu} must be a positive definite matrix. If this condition is satisfied, then $|H_{uu}| > 0$ and the inverse H_{uu}^{-1} exists (as we have heretofore assumed). If H_{uu} is only positive semi-definite at all points on some arc of the extremal, then $|H_{uu}| = 0$ over this arc and the inverse matrix fails to exist; i.e. H_{uu} is *singular*. In these circumstances, the arc is also said to be *singular* and our sufficiency condition fails to decide whether the extremal is optimal.

The necessary conditions $\partial H/\partial u_r = 0$ can only be solved for the control variables u_r in terms of the state variables x_i and multipliers λ_i provided the Jacobian $|H_{uu}|$ is non-vanishing. Along a singular arc, therefore, these necessary conditions (by themselves) fail to provide an optimal control law.

Singular arcs most frequently arise in problems where the Hamiltonian depends linearly on the control variables. Then $\partial^2 H/\partial u_r \partial u_s$ all vanish identically and all optimal trajectories are singular. A linear control system having a quadratic performance index belongs to this category of systems with singular extremals if the integrand of the index contains no terms of the second degree in the control variables.

PROBLEM 34. In the 4-terminal network described in Problem 11, the output quantity x is to be changed from an initial value x_0 at $t = 0$ to a final value x_1 at $t = T$, in such a way that

$$C = \int_0^T x^2 \, dt$$

is minimized (T is prescribed).

Solution: The Hamiltonian is $H = x^2 + \lambda(-2x + u)$ and Hamilton's equations are accordingly

$$\dot{x} = -2x + u, \qquad \dot{\lambda} = -2x + 2\lambda, \qquad \lambda = 0.$$

These equations only possess the solution $x = \lambda = u = 0$, determining a singular extremal.

Suppose the system is subjected to an impulsive input $u = -x_0\delta(t)$ at $t = 0$, $\delta(t)$ being the Dirac unit impulse function. Integrating the state equation over the small duration τ of this impulse, we calculate that the effect of the impulse is to cause an increase in x of amount

$$\Delta x = \int_0^\tau (-2x - x_0\delta(t)) \, dt = -x_0.$$

(Note: x remains finite during the short interval $(0, \tau)$ and hence makes negligible contribution to the integral.) Thus, x is reduced to the value zero.

Now suppose that the control u is maintained at the value 0 until the instant $t = T$; during this period the system will remain quiescent, its behaviour being described by the singular extremal $x = 0$.

At $t = T$, an impulse $x_1\delta(t - T)$ is applied, with the result that the value of x is increased from 0 to x_1, thus satisfying the end constraint.

Evidently, the value of C associated with this two-impulse control is zero and, therefore, C has been minimized. ●

In the general case of a system possessing singular extremal arcs, the optimal control is frequently of this type, namely, impulse-singular arc-impulse.

PROBLEM 35. A linear system is governed by the equations $\dot{x} = y + u$, $\dot{y} = -x + u$, u being the control quantity. It is to be moved from the initial state $(1, 2)$ at $t = 0$ to the final state $(0, 0)$ at $t = T$

in such a way that

$$C = \int_0^T x^2 \, dt$$

is minimized. Calculate the optimal control.

Solution: In this case

$$H = x^2 + \lambda(y + u) + \mu(-x + u).$$

Hamilton's equations are

$$\dot\lambda = -2x + \mu, \quad \dot\mu = -\lambda, \quad \lambda + \mu = 0.$$

Eliminating μ, we find $\dot\lambda = -2x - \lambda$, $\dot\lambda = \lambda$. It follows that $\lambda = -x$ and $\dot x = x$. Hence

$$x = -\lambda = \mu = Pe^t,$$

where P is constant. The state equations now show that

$$y = Qe^{-t}, \quad u = Pe^t - Qe^{-t},$$

where Q is a second constant.

The equations just derived are valid along the singular extremal arcs. Thus, $xy = PQ$ and, in the state plane, the singular arcs are seen to be rectangular hyperbolae.

Optimal transfer between the initial and final states is effected by (i) application of an initial impulse $u = -J\delta(t)$ which moves the system from the state $A(1, 2)$ to the state $B(1 - J, 2 - J)$ (Fig. 5.2), (ii) motion along the singular arc BC and (iii) application of a final impulse $u = -K\delta(t - T)$ which moves the system from the state $C(K, K)$ to the final state $(0, 0)$. Note that on the singular arc BC, x and y are both positive and, hence, P and Q are positive constants; it follows from the equations above that x increases and y decreases as time elapses and the sense of description of this arc must be the one indicated. Note, also, that the two impulses make no contribution to the cost integral C.

Taking the equation of the hyperbolic arc BC to be $xy = PQ$, it is found that $J = [3 - (1 + 4PQ)^{\frac{1}{2}}]/2$ and $K = (PQ)^{\frac{1}{2}}$. The coordinates of the points B, C can now be expressed in terms of PQ. But, $t = 0$ at B and, hence, this point has coordinates (P, Q); thus,

$$2P + 1 = (1 + 4PQ)^{\frac{1}{2}}.$$

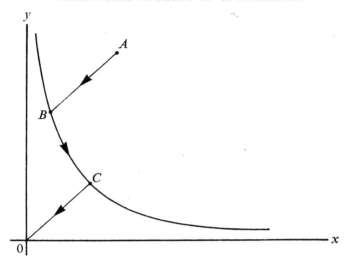

Fig. 5.2

At C, $t = T$ and its coordinates are therefore (Pe^T, Qe^{-T}); thus

$$Pe^T = (PQ)^{\frac{1}{2}}.$$

It now follows that

$$P = \frac{1}{e^{2T} - 1}, \qquad Q = \frac{e^{2T}}{e^{2T} - 1}.$$

Between the two impulses, the optimal control is accordingly determined by the equation

$$u = -\sinh(T - t)/\sinh T.$$

The optimal cost is given by

$$C_{\text{opt}} = \int_0^T P^2 e^{2t}\, dt = \frac{1}{2(e^{2T} - 1)}.$$

PROBLEM 36. Calculate the singular arcs for a rocket moving in a plane in an inverse square law gravitational field, the exhaust velocity being constant and it being required to minimize the propellent expenditure.

Solution: We shall employ the notation introduced in Section 2.8 and Fig. 2.3. The acceleration generated by the motor thrust is given by $f = cm/M$, where c is the exhaust velocity, M is the rocket mass and $m(= -\dot{M})$ is the rate of expenditure of propellent. The state equations accordingly take the form

$$\dot{r} = u, \qquad \dot{\theta} = v/r \tag{5.6.1}$$

$$\dot{u} = \frac{v^2}{r} - \frac{\gamma}{r^2} + \frac{cm}{M} \sin \phi \tag{5.6.2}$$

$$\dot{v} = -\frac{uv}{r} + \frac{cm}{M} \cos \phi \tag{5.6.3}$$

$$\dot{M} = -m \tag{5.6.4}$$

γ/r^2 being the gravitational acceleration at distance r from the centre of attraction O. Thus, (r, θ, u, v, M) are the components of the state vector and (m, ϕ) are the components of the control vector.

The cost function is $C = M_1$ (the final rocket mass) and this must, of course, be maximized. We shall assume, as usual, that the state vector is known at $t = t_0$ and that the components (r, θ, u, v) are known at $t = t_1$. The time of arrival t_1 may be predetermined or may be open to choice.

The Hamiltonian is constructed in the form

$$H = \lambda_r u + \lambda_\theta \frac{v}{r} + \lambda_u \left(\frac{v^2}{r} - \frac{\gamma}{r^2} + \frac{cm}{M} \sin \phi \right)$$
$$+ \lambda_v \left(-\frac{uv}{r} + \frac{cm}{M} \cos \phi \right) - \lambda_M m \tag{5.6.5}$$

and this leads to the Hamilton equations

$$\dot{\lambda}_r = \lambda_\theta \frac{v}{r^2} + \lambda_u \left(\frac{v^2}{r^2} - \frac{2\gamma}{r^3} \right) - \lambda_v \frac{uv}{r^2}, \tag{5.6.6}$$

$$\dot{\lambda}_\theta = 0, \tag{5.6.7}$$

$$\dot{\lambda}_u = -\lambda_r + \lambda_v \frac{v}{r}, \tag{5.6.8}$$

$$\dot{\lambda}_v = -\lambda_\theta \frac{1}{r} - \lambda_u \frac{2v}{r} + \lambda_v \frac{u}{r}, \tag{5.6.9}$$

$$\dot{\lambda}_M = \frac{cm}{M^2} (\lambda_u \sin \phi + \lambda_v \cos \phi) \tag{5.6.10}$$

together with the optimal control equations

$$\frac{\partial H}{\partial m} = \frac{c}{M}(\lambda_u \sin \phi + \lambda_v \cos \phi) - \lambda_M = 0, \quad (5.6.11)$$

$$\frac{\partial H}{\partial \phi} = \frac{cm}{M}(\lambda_u \cos \phi - \lambda_v \sin \phi) = 0. \quad (5.6.12)$$

We note that

$$\frac{\partial^2 H}{\partial m^2} = 0, \quad \frac{\partial^2 H}{\partial m \partial \phi} = \frac{c}{M}(\lambda_u \cos \phi - \lambda_v \sin \phi),$$

$$\frac{\partial^2 H}{\partial \phi^2} = -\frac{cm}{M}(\lambda_u \sin \phi + \lambda_v \cos \phi).$$

It follows from equation (5.6.12) that, along an extremal for which $m \neq 0$, $\partial^2 H/\partial m \partial \phi = 0$ and this implies that this type of extremal is always singular. If m is subject to inequality constraints, as assumed in Section 4.5 (see expression (4.5.6)), two other types of non-singular extremal will occur, namely, arcs of maximum thrust and arcs of null thrust. Attention is being confined to the singular intermediate thrust arcs in this problem.

It follows from equation (5.6.12) that

$$\lambda_u = \lambda \sin \phi, \quad \lambda_v = \lambda \cos \phi. \quad (5.6.13)$$

Substituting for λ_u, λ_v into equations (5.6.10), (5.6.11), it is found that

$$\dot{\lambda}_M = cm\lambda/M^2, \quad \lambda_M = c\lambda/M. \quad (5.6.14)$$

For these equations to be consistent, it is necessary that λ be constant. λ_u and λ_v are the components of the primer vector introduced in Section 4.5 and, as in that section, we have therefore proved that, along a singular IT arc, the magnitude of the primer is constant.

Substituting for λ_u, λ_v from equations (5.6.13) into equation (5.6.9) and putting $\lambda_\theta = -A\lambda$ (a constant), we derive the equation

$$-u \cos \phi + 2v \sin \phi - r\dot{\phi} \sin \phi = A. \quad (5.6.15)$$

Since H does not depend explicitly on the time, a first integral $H = B$ is also available. After substituting for λ_u, λ_v from equations (5.6.13) and making use of the second of equations (5.6.14), this is found to reduce to

$$\left(\lambda_r - \lambda \frac{v}{r} \cos \phi\right) u + \lambda \left(\frac{v^2}{r} - \frac{\gamma}{r^2}\right) \sin \phi - \lambda A \frac{v}{r} = B. \quad (5.6.16)$$

But, equation (5.6.8) shows that

$$\lambda \dot{\phi} \cos \phi = -\lambda_r + \lambda \frac{v}{r} \cos \phi. \qquad (5.6.17)$$

Hence, the first integral (5.6.16) further reduces to

$$u\dot{\phi} \cos \phi + \frac{Av}{r} - \left(\frac{v^2}{r} - \frac{\gamma}{r^2}\right)\sin \phi = D, \qquad (5.6.18)$$

where $D = -B/\lambda$. Substitution for A from equation (5.6.15) into this last equation, enables us to write this integral in the form

$$w\left(\frac{v}{r} - \dot{\phi}\right) = D - \frac{\gamma}{r^2}\sin \phi, \qquad (5.6.19)$$

where

$$w = v \sin \phi - u \cos \phi, \qquad (5.6.20)$$

i.e. w is the velocity component in the direction of the motor thrust. The integral (5.6.15) can also be expressed conveniently in terms of w, thus:

$$w + \left(\frac{v}{r} - \dot{\phi}\right) r \sin \phi = A. \qquad (5.6.21)$$

Yet another first order equation can be derived as follows: Differentiating equation (5.6.17), substituting from equation (5.6.6) for λ_r and making use of equations (5.6.13), (5.6.15), we first obtain the result

$$r(\ddot{\phi} \cos \phi - \dot{\phi}^2 \sin \phi) = \left(\frac{v^2}{r} + \frac{2\gamma}{r^2}\right)\sin \phi + \dot{v} \cos \phi$$
$$- \frac{uv}{r} \cos \phi - 2v\dot{\phi} \sin \phi. \qquad (5.6.22)$$

Then, by differentiating equation (5.6.15), it follows that

$$r(\ddot{\phi} \sin \phi + \dot{\phi}^2 \cos \phi) = -\dot{u} \cos \phi + 2\dot{v} \sin \phi + 2v\dot{\phi} \cos \phi. \qquad (5.6.23)$$

Eliminating $\ddot{\phi}$ between the last two equations, it will be found that

$$r\dot{\phi}^2 = (\dot{v} \sin \phi - \dot{u} \cos \phi)\cos \phi + 2v\dot{\phi}$$
$$- \left(\frac{v^2}{r} + \frac{2\gamma}{r^2}\right)\sin^2 \phi + \frac{uv}{r} \sin \phi \cos \phi. \qquad (5.6.24)$$

Making use of equations (5.6.2), (5.6.3), this equation finally reduces to the form

$$\left(\frac{v}{r} - \dot\phi\right)^2 = \frac{\gamma}{r^3}(1 - 3\sin^2\phi). \tag{5.6.25}$$

We now eliminate $(v/r - \dot\phi)$ between equations (5.6.19), (5.6.21), (5.6.25), to yield the two equations

$$w(A - w) = \left(Dr - \frac{\gamma}{r}\sin\phi\right)\sin\phi, \tag{5.6.26}$$

$$w^2(1 - 3\sin^2\phi) = \frac{r}{\gamma}\left(Dr - \frac{\gamma}{r}\sin\phi\right)^2. \tag{5.6.27}$$

These equations can be solved for w and r as functions of ϕ and, then, all the remaining variables become expressible in terms of the parameter ϕ.

Thus, equation (5.6.20) is equivalent to the equation

$$w = -\frac{d}{dt}(r\cos\phi) + r(\dot\theta - \dot\phi)\sin\phi \tag{5.6.28}$$

and, by equation (5.6.21), this implies that

$$\frac{d}{dt}(r\cos\phi) = A - 2w. \tag{5.6.29}$$

Hence

$$(A - 2w)\frac{dt}{d\phi} = \frac{d}{d\phi}(r\cos\phi) \tag{5.6.30}$$

and an integration yields t as a function of ϕ. The inverse relationship $\phi = \phi(t)$ is the control law for ϕ. It now follows from equation (5.6.21) that

$$\frac{d\theta}{d\phi} = 1 + \frac{A - w}{r\sin\phi}\frac{dt}{d\phi}. \tag{5.6.31}$$

Integrating for θ in terms of ϕ, we are now provided with two parametric equations for the singular arc in the form $r = r(\phi)$, $\theta = \theta(\phi)$. The remaining variables can be expressed in terms of ϕ by use of the equations of motion (5.6.1)–(5.6.3). In particular, the control law for m can now be found.

In particular, if the transit time is not predetermined, then $D = 0$ and it will be found that the parametric equations for the family of singular arcs are

$$r = \frac{a \sin^6 \phi}{1 - 3 \sin^2 \phi}, \qquad \theta = \theta_0 - 4\phi - 3 \cot \phi, \qquad (5.6.32)$$

a, θ_0 being constants. These equations determine a family of spirals. It has been proved by H. M. Robbins and by R. E. Kopp and H. G. Moyer[*] that these spirals are non-optimal. ●

Exercises 5

1. In Problem 30, replace the cost integral by

$$C = \int_0^2 x^{\frac{1}{2}} \sec \theta \, dt$$

and suppose the system is to be transferred from the state $x = 2$ at $t = 0$ to the state $x = 5$ at $t = 2$. Show that there are two possible extremals, but that the one which satisfies Jacobi's condition is given by $x = \frac{1}{4}t^2 + t + 2$. Deduce that $C_{\min} = 20/3$.

2. In Problem 30, replace the cost integral by

$$C = \int_0^1 \frac{1}{x} \sec \theta \, dt$$

and suppose the system is to be transferred from the state $x = 1$ at $t = 0$ to the state $x = \sqrt{2}$ at $t = 1$. Show that only one possible extremal exists and that this is given by $x = \sqrt{(1 + 2t - t^2)}$. Show that Jacobi's condition is satisfied and deduce that

$$C_{\min} = \log(\sqrt{2} + 1).$$

3. A system's state equation is $\dot{x} = u$. It is to be transferred from the state $x = 0$ at $t = 0$ to the state $x = 2$ at $t = 1$. The cost to be minimized is given by

$$C = \int_0^1 (x + \tfrac{1}{4}u^2) \, dt.$$

---

[*] Robbins, H. M., 'Optimality of Intermediate-Thrust Arcs of Rocket Trajectories', *AIAA Journ.*, Vol. 3, No. 6, pp. 1094–1098. Kopp, R.E. and Moyer, H. G., 'Necessary Conditions for Singular Extremals', *AIAA Journ.*, Vol. 3, No. 8, pp. 1439–1444.

Show that the only possible optimal motion is given by $x = t^2 + t$ and that Jacobi's condition is satisfied. Obtain the minimal cost. (Answer: 23/12.)

4. In the last exercise, replace the cost of integral by

$$C = \int_0^{\frac{1}{4}\pi} \sqrt{(x^2 + u^2)}\, dt$$

and assume that transfer is required from the state $x = 1$ at $t = 0$ to the state $x = \sqrt{2}$ at $t = \frac{1}{4}\pi$. Show, again, that the optimal motion is unique and that the optimal control is given by $u = \sec t \tan t$. Show that Jacobi's condition is satisfied and that $C_{\min}$ is unity.

5. The state equation for a second order system is $\ddot{x} = u$. It is to be transferred from the state $x = x_0$, $\dot{x} = y_0$ at $t = t_0$ to the state $x = x_1$, $\dot{x} = y_1$ at $t = t_1$ with minimization of the cost

$$C = \int_{t_0}^{t_1} u^2\, dt.$$

Show that for optimal control u is linear in t and that Jacobi's condition is satisfied.

6. For Exercise 1, show that the accessory Riccati equation on the extremal satisfying the Jacobi condition can be expressed in the form

$$\frac{dK}{d\theta} = 2K^2 \sec^4 \theta + 2K \tan \theta \sec^2 \theta + \tfrac{1}{2} \sec^2 \theta.$$

Verify that $K = -\tfrac{1}{2}\cot \theta$ is a solution of this equation. Obtain the general solution and verify that a solution which is finite over the whole of the extremal arc can be found.

7. For Exercise 3, obtain the accessory Riccati equation on any extremal and deduce that its general solution is $K = 1/(A - 2t)$, where A is an arbitrary constant. Deduce that every extremal satisfies the Jacobi condition.

8. Show that the accessory Riccati equation on the extremal specified in Exercise 4 takes the form

$$\frac{dK}{dt} = K^2 \sec^4 t - 2K \tan t.$$

Putting $K = 1/J$, obtain the general solution of this equation and deduce that a solution can be found which is finite over the whole extremal. Show, however, that no such solution exists if the extremal arc is extended from $t = \frac{1}{4}\pi$ to $t = \pi$.

9. Show that the general symmetric solution of the accessory Riccati equation for any extremal in Exercise 5 can be expressed in the form

$$K = 2(\alpha - t)^{-1} \begin{bmatrix} \alpha\gamma + \beta^2 - \gamma t & \beta \\ \beta & 1 \end{bmatrix},$$

where α, β, γ are constants. Deduce that a solution exists for K which is finite over every extremal.

BIBLIOGRAPHY

BLISS, G. A., Lectures on the Calculus of Variations, University of Chicago Press, 1961.

BRYSON, A. E., HO, Y. C., Applied Optimal Control, Ginn, 1969.

GUMOWSKI, I., MIRA, C., Optimization in Control Theory and Practice, Cambridge University Press, 1968.

HESTENES, M., Calculus of Variations and Optimal Control Theory, Wiley, 1966.

LAWDEN, D. F., Optimal Trajectories for Space Navigation, Butterworth, 1963.

LEE, E. B., MARKUS, L., Foundations of Optimal Control Theory, Wiley, 1967.

LEITMANN, G. (ed.) Optimization Techniques with Applications to Aerospace Systems, Academic Press, 1962.

LEITMANN, G., An Introduction to Optimal Control, McGraw Hill, 1966.

PARS, L. A., An Introduction to the Calculus of Variations, Heinemann, 1962.

PONTRYAGIN, L. S., BOLTYANSKII, V. G., GAMKRELIDZE, R. V., MISHCHENKO, E. F. The Mathematical Theory of Optimal Processes, Interscience, 1962.

INDEX

Accessory equations   127, 133
Accessory equations, solution of   130
Accessory problem   126
Adjoint equations   9, 39, 40, 65, 82, 83
Admissible controls   62
Admissible point   7
Admissible trajectories   37
Attainability, set of   63
Autonomous system   26, 33

Bang-bang control   122
Basset, A. B.   75
Bliss, G. A.   37, 153
Boltyanskii, V. G.   153
Bryson, A. E.   153

Cayley-Hamilton theorem   31
Clebsch condition   121, 138, 139
Conjugate points   127, 135
Constraints, equality   7
Constraints, inequality   14, 111, 117
Constraints, integral   86, 106
Constraints, non-differential   81, 106, 111, 117
Controllable, system completely   62
Control space   2
Control trajectory   24
Control variables   1, 24
Control variables, discontinuity of   35, 36, 90, 92
Control vector   2, 8
Cost index   1, 36

Definite form   4
Duffing oscillator   61
Dynamic systems   2, 24

Eigenvalues of matrix   4, 32, 35, 53
Elastic string   80
Embedding of optimal trajectory   37
Euler's equation   79
Euler's equation, integrals of   80
Extremal   134

Fermat's principle   76
Final instant   25, 62
Final state   62
Final state, constraints on   62
Four terminal network   41, 68, 86, 98, 144
Functional   2

Gamkrelidze, R. V.   153
Gumowski, I.   153

Hamiltonian   9, 39, 65, 82, 83, 86, 111, 127
Hamilton-Jacobi equation   105
Hamilton's equations   40
Hamilton's equations, integral of   72
Hamilton's principle   84
Hestenes, M.   153
Ho, Y. C.   153
Hypocycloid   75

Initial instant   25
Initial state   25, 103
Input variables   24

Jacobi's condition   128, 138, 139
Junction points   119

Kopp, R. E.   151
Kronecker delta   26

INDEX

Lagrangian 84
Lawden, D. F. 120, 153
Lee, E. B. 153
Leitmann, G. 153
Linear control system 26, 33, 47, 94, 128
Linear programming 18
Linear programming, dual problem 18, 21

Markus, L. 153
Matrices, series of 30
Matrix, exponential function of 31
Minimum, local 5
Minimum, necessary conditions for 3, 123
Minimum, sufficient conditions for 5, 138
Mira, C. 153
Mishchenko, E. F. 153
Moyer, H. G. 151
Multipliers, Lagrange 9, 13, 39, 65, 82

Neighbourhood 4
Numerical integration 50, 56, 97

Optimal control law 42, 50
Optimal control, necessary conditions for 37, 40, 66
Optimal cost, formula for 13, 100
Optimal cost function 103
Optimal cost function, derivatives of 108
Optimal point 3
Optimal trajectories, field of 103
Optimization 1
Orbit, escape from 53
Output variables 26

Pars, L. A. 153
Penalty index 1
Performance Index 1, 36
Pontryagin, L. S. 153
Pontryagin's principle 106, 111, 113, 115, 118, 120
Primer vector 118, 148

Quadratic cost function 12, 23, 47, 94, 128

Response, optimal 1, 36
Response, system's 1
Riccati equation 45, 49
Riccati equation, accessory 134, 138
Riccati equation, steady state 52
Robbins, H. M. 151
Rocket, singular arcs for a 146
Rocket trajectory, conditions for optimal 120
Rocket trajectory, optimal 116
Rocket vehicle 42, 50, 53, 69 102, 116, 122

Satellite, artificial 45
Semi-definite form 4
Singular arcs 143
Spirals, non-optimal 151
Stable system 36
State equations 24
State equations, canonical form 25
State space, augmented 103
State trajectory 24
State transition matrix 27, 33
State transition matrix, multiplicative property of 27
State variables 24
State vector 8
Static systems 2
Summation convention 3
Switching function 118
Switching point 113, 115

Throw-off angle 59
Thrust, arcs of intermediate rocket 118, 148
Thrust, arcs of maximum rocket 118
Thrust, arcs of null rocket 118
Transversality condition 80
Tunnel, earth 72

Unconstrained control 3
Unstable system 36

Variation, equations of 38, 64, 124
Variation of control variables 37, 63, 124
Variation of cost, first 124
Variation of cost, second 124, 126

Variation of state variables 38, 63, 124
Variations, calculus of 78

Weierstrass condition 120

A CATALOG OF SELECTED
DOVER BOOKS
IN SCIENCE AND MATHEMATICS

CATALOG OF DOVER BOOKS

Astronomy

BURNHAM'S CELESTIAL HANDBOOK, Robert Burnham, Jr. Thorough guide to the stars beyond our solar system. Exhaustive treatment. Alphabetical by constellation: Andromeda to Cetus in Vol. 1; Chamaeleon to Orion in Vol. 2; and Pavo to Vulpecula in Vol. 3. Hundreds of illustrations. Index in Vol. 3. 2,000pp. 6⅛ x 9¼.
Vol. I: 0-486-23567-X
Vol. II: 0-486-23568-8
Vol. III: 0-486-23673-0

EXPLORING THE MOON THROUGH BINOCULARS AND SMALL TELESCOPES, Ernest H. Cherrington, Jr. Informative, profusely illustrated guide to locating and identifying craters, rills, seas, mountains, other lunar features. Newly revised and updated with special section of new photos. Over 100 photos and diagrams. 240pp. 8¼ x 11. 0-486-24491-1

THE EXTRATERRESTRIAL LIFE DEBATE, 1750–1900, Michael J. Crowe. First detailed, scholarly study in English of the many ideas that developed from 1750 to 1900 regarding the existence of intelligent extraterrestrial life. Examines ideas of Kant, Herschel, Voltaire, Percival Lowell, many other scientists and thinkers. 16 illustrations. 704pp. 5⅜ x 8½. 0-486-40675-X

THEORIES OF THE WORLD FROM ANTIQUITY TO THE COPERNICAN REVOLUTION, Michael J. Crowe. Newly revised edition of an accessible, enlightening book recreates the change from an earth-centered to a sun-centered conception of the solar system. 242pp. 5⅜ x 8½. 0-486-41444-2

A HISTORY OF ASTRONOMY, A. Pannekoek. Well-balanced, carefully reasoned study covers such topics as Ptolemaic theory, work of Copernicus, Kepler, Newton, Eddington's work on stars, much more. Illustrated. References. 521pp. 5⅜ x 8½.
0-486-65994-1

A COMPLETE MANUAL OF AMATEUR ASTRONOMY: TOOLS AND TECHNIQUES FOR ASTRONOMICAL OBSERVATIONS, P. Clay Sherrod with Thomas L. Koed. Concise, highly readable book discusses: selecting, setting up and maintaining a telescope; amateur studies of the sun; lunar topography and occultations; observations of Mars, Jupiter, Saturn, the minor planets and the stars; an introduction to photoelectric photometry; more. 1981 ed. 124 figures. 25 halftones. 37 tables. 335pp. 6½ x 9¼. 0-486-40675-X

AMATEUR ASTRONOMER'S HANDBOOK, J. B. Sidgwick. Timeless, comprehensive coverage of telescopes, mirrors, lenses, mountings, telescope drives, micrometers, spectroscopes, more. 189 illustrations. 576pp. 5⅜ x 8¼. (Available in U.S. only.)
0-486-24034-7

STARS AND RELATIVITY, Ya. B. Zel'dovich and I. D. Novikov. Vol. 1 of *Relativistic Astrophysics* by famed Russian scientists. General relativity, properties of matter under astrophysical conditions, stars, and stellar systems. Deep physical insights, clear presentation. 1971 edition. References. 544pp. 5⅜ x 8¼. 0-486-69424-0

Chemistry

THE SCEPTICAL CHYMIST: THE CLASSIC 1661 TEXT, Robert Boyle. Boyle defines the term "element," asserting that all natural phenomena can be explained by the motion and organization of primary particles. 1911 ed. viii+232pp. 5⅜ x 8½.
0-486-42825-7

RADIOACTIVE SUBSTANCES, Marie Curie. Here is the celebrated scientist's doctoral thesis, the prelude to her receipt of the 1903 Nobel Prize. Curie discusses establishing atomic character of radioactivity found in compounds of uranium and thorium; extraction from pitchblende of polonium and radium; isolation of pure radium chloride; determination of atomic weight of radium; plus electric, photographic, luminous, heat, color effects of radioactivity. ii+94pp. 5⅜ x 8½. 0-486-42550-9

CHEMICAL MAGIC, Leonard A. Ford. Second Edition, Revised by E. Winston Grundmeier. Over 100 unusual stunts demonstrating cold fire, dust explosions, much more. Text explains scientific principles and stresses safety precautions. 128pp. 5⅜ x 8½.
0-486-67628-5

THE DEVELOPMENT OF MODERN CHEMISTRY, Aaron J. Ihde. Authoritative history of chemistry from ancient Greek theory to 20th-century innovation. Covers major chemists and their discoveries. 209 illustrations. 14 tables. Bibliographies. Indices. Appendices. 851pp. 5⅜ x 8½. 0-486-64235-6

CATALYSIS IN CHEMISTRY AND ENZYMOLOGY, William P. Jencks. Exceptionally clear coverage of mechanisms for catalysis, forces in aqueous solution, carbonyl- and acyl-group reactions, practical kinetics, more. 864pp. 5⅜ x 8½.
0-486-65460-5

ELEMENTS OF CHEMISTRY, Antoine Lavoisier. Monumental classic by founder of modern chemistry in remarkable reprint of rare 1790 Kerr translation. A must for every student of chemistry or the history of science. 539pp. 5⅜ x 8½. 0-486-64624-6

THE HISTORICAL BACKGROUND OF CHEMISTRY, Henry M. Leicester. Evolution of ideas, not individual biography. Concentrates on formulation of a coherent set of chemical laws. 260pp. 5⅜ x 8½.
0-486-61053-5

A SHORT HISTORY OF CHEMISTRY, J. R. Partington. Classic exposition explores origins of chemistry, alchemy, early medical chemistry, nature of atmosphere, theory of valency, laws and structure of atomic theory, much more. 428pp. 5⅜ x 8½. (Available in U.S. only.)
0-486-65977-1

GENERAL CHEMISTRY, Linus Pauling. Revised 3rd edition of classic first-year text by Nobel laureate. Atomic and molecular structure, quantum mechanics, statistical mechanics, thermodynamics correlated with descriptive chemistry. Problems. 992pp. 5⅜ x 8½.
0-486-65622-5

FROM ALCHEMY TO CHEMISTRY, John Read. Broad, humanistic treatment focuses on great figures of chemistry and ideas that revolutionized the science. 50 illustrations. 240pp. 5⅜ x 8½.
0-486-28690-8

CATALOG OF DOVER BOOKS

Mathematics

FUNCTIONAL ANALYSIS (Second Corrected Edition), George Bachman and Lawrence Narici. Excellent treatment of subject geared toward students with background in linear algebra, advanced calculus, physics and engineering. Text covers introduction to inner-product spaces, normed, metric spaces, and topological spaces; complete orthonormal sets, the Hahn-Banach Theorem and its consequences, and many other related subjects. 1966 ed. 544pp. 6⅛ x 9¼. 0-486-40251-7

ASYMPTOTIC EXPANSIONS OF INTEGRALS, Norman Bleistein & Richard A. Handelsman. Best introduction to important field with applications in a variety of scientific disciplines. New preface. Problems. Diagrams. Tables. Bibliography. Index. 448pp. 5⅜ x 8½. 0-486-65082-0

VECTOR AND TENSOR ANALYSIS WITH APPLICATIONS, A. I. Borisenko and I. E. Tarapov. Concise introduction. Worked-out problems, solutions, exercises. 257pp. 5⅜ x 8¼. 0-486-63833-2

AN INTRODUCTION TO ORDINARY DIFFERENTIAL EQUATIONS, Earl A. Coddington. A thorough and systematic first course in elementary differential equations for undergraduates in mathematics and science, with many exercises and problems (with answers). Index. 304pp. 5⅜ x 8½. 0-486-65942-9

FOURIER SERIES AND ORTHOGONAL FUNCTIONS, Harry F. Davis. An incisive text combining theory and practical example to introduce Fourier series, orthogonal functions and applications of the Fourier method to boundary-value problems. 570 exercises. Answers and notes. 416pp. 5⅜ x 8½. 0-486-65973-9

COMPUTABILITY AND UNSOLVABILITY, Martin Davis. Classic graduate-level introduction to theory of computability, usually referred to as theory of recurrent functions. New preface and appendix. 288pp. 5⅜ x 8½. 0-486-61471-9

ASYMPTOTIC METHODS IN ANALYSIS, N. G. de Bruijn. An inexpensive, comprehensive guide to asymptotic methods–the pioneering work that teaches by explaining worked examples in detail. Index. 224pp. 5⅜ x 8½ 0-486-64221-6

APPLIED COMPLEX VARIABLES, John W. Dettman. Step-by-step coverage of fundamentals of analytic function theory–plus lucid exposition of five important applications: Potential Theory; Ordinary Differential Equations; Fourier Transforms; Laplace Transforms; Asymptotic Expansions. 66 figures. Exercises at chapter ends. 512pp. 5⅜ x 8½. 0-486-64670-X

INTRODUCTION TO LINEAR ALGEBRA AND DIFFERENTIAL EQUATIONS, John W. Dettman. Excellent text covers complex numbers, determinants, orthonormal bases, Laplace transforms, much more. Exercises with solutions. Undergraduate level. 416pp. 5⅜ x 8½. 0-486-65191-6

RIEMANN'S ZETA FUNCTION, H. M. Edwards. Superb, high-level study of landmark 1859 publication entitled "On the Number of Primes Less Than a Given Magnitude" traces developments in mathematical theory that it inspired. xiv+315pp. 5⅜ x 8½. 0-486-41740-9

CATALOG OF DOVER BOOKS

CALCULUS OF VARIATIONS WITH APPLICATIONS, George M. Ewing. Applications-oriented introduction to variational theory develops insight and promotes understanding of specialized books, research papers. Suitable for advanced undergraduate/graduate students as primary, supplementary text. 352pp. 5⅜ x 8½.
0-486-64856-7

COMPLEX VARIABLES, Francis J. Flanigan. Unusual approach, delaying complex algebra till harmonic functions have been analyzed from real variable viewpoint. Includes problems with answers. 364pp. 5⅜ x 8½.
0-486-61388-7

AN INTRODUCTION TO THE CALCULUS OF VARIATIONS, Charles Fox. Graduate-level text covers variations of an integral, isoperimetrical problems, least action, special relativity, approximations, more. References. 279pp. 5⅜ x 8½.
0-486-65499-0

COUNTEREXAMPLES IN ANALYSIS, Bernard R. Gelbaum and John M. H. Olmsted. These counterexamples deal mostly with the part of analysis known as "real variables." The first half covers the real number system, and the second half encompasses higher dimensions. 1962 edition. xxiv+198pp. 5⅜ x 8½. 0-486-42875-3

CATASTROPHE THEORY FOR SCIENTISTS AND ENGINEERS, Robert Gilmore. Advanced-level treatment describes mathematics of theory grounded in the work of Poincaré, R. Thom, other mathematicians. Also important applications to problems in mathematics, physics, chemistry and engineering. 1981 edition. References. 28 tables. 397 black-and-white illustrations. xvii + 666pp. 6⅛ x 9¼.
0-486-67539-4

INTRODUCTION TO DIFFERENCE EQUATIONS, Samuel Goldberg. Exceptionally clear exposition of important discipline with applications to sociology, psychology, economics. Many illustrative examples; over 250 problems. 260pp. 5⅜ x 8½.
0-486-65084-7

NUMERICAL METHODS FOR SCIENTISTS AND ENGINEERS, Richard Hamming. Classic text stresses frequency approach in coverage of algorithms, polynomial approximation, Fourier approximation, exponential approximation, other topics. Revised and enlarged 2nd edition. 721pp. 5⅜ x 8½. 0-486-65241-6

INTRODUCTION TO NUMERICAL ANALYSIS (2nd Edition), F. B. Hildebrand. Classic, fundamental treatment covers computation, approximation, interpolation, numerical differentiation and integration, other topics. 150 new problems. 669pp. 5⅜ x 8½.
0-486-65363-3

THREE PEARLS OF NUMBER THEORY, A. Y. Khinchin. Three compelling puzzles require proof of a basic law governing the world of numbers. Challenges concern van der Waerden's theorem, the Landau-Schnirelmann hypothesis and Mann's theorem, and a solution to Waring's problem. Solutions included. 64pp. 5⅜ x 8½.
0-486-40026-3

THE PHILOSOPHY OF MATHEMATICS: AN INTRODUCTORY ESSAY, Stephan Körner. Surveys the views of Plato, Aristotle, Leibniz & Kant concerning propositions and theories of applied and pure mathematics. Introduction. Two appendices. Index. 198pp. 5⅜ x 8½.
0-486-25048-2

CATALOG OF DOVER BOOKS

INTRODUCTORY REAL ANALYSIS, A.N. Kolmogorov, S. V. Fomin. Translated by Richard A. Silverman. Self-contained, evenly paced introduction to real and functional analysis. Some 350 problems. 403pp. 5⅜ x 8½. 0-486-61226-0

APPLIED ANALYSIS, Cornelius Lanczos. Classic work on analysis and design of finite processes for approximating solution of analytical problems. Algebraic equations, matrices, harmonic analysis, quadrature methods, much more. 559pp. 5⅜ x 8½. 0-486-65656-X

AN INTRODUCTION TO ALGEBRAIC STRUCTURES, Joseph Landin. Superb self-contained text covers "abstract algebra": sets and numbers, theory of groups, theory of rings, much more. Numerous well-chosen examples, exercises. 247pp. 5⅜ x 8½. 0-486-65940-2

QUALITATIVE THEORY OF DIFFERENTIAL EQUATIONS, V. V. Nemytskii and V.V. Stepanov. Classic graduate-level text by two prominent Soviet mathematicians covers classical differential equations as well as topological dynamics and ergodic theory. Bibliographies. 523pp. 5⅜ x 8½. 0-486-65954-2

THEORY OF MATRICES, Sam Perlis. Outstanding text covering rank, nonsingularity and inverses in connection with the development of canonical matrices under the relation of equivalence, and without the intervention of determinants. Includes exercises. 237pp. 5⅜ x 8½. 0-486-66810-X

INTRODUCTION TO ANALYSIS, Maxwell Rosenlicht. Unusually clear, accessible coverage of set theory, real number system, metric spaces, continuous functions, Riemann integration, multiple integrals, more. Wide range of problems. Undergraduate level. Bibliography. 254pp. 5⅜ x 8½. 0-486-65038-3

MODERN NONLINEAR EQUATIONS, Thomas L. Saaty. Emphasizes practical solution of problems; covers seven types of equations. ". . . a welcome contribution to the existing literature...."–*Math Reviews*. 490pp. 5⅜ x 8½. 0-486-64232-1

MATRICES AND LINEAR ALGEBRA, Hans Schneider and George Phillip Barker. Basic textbook covers theory of matrices and its applications to systems of linear equations and related topics such as determinants, eigenvalues and differential equations. Numerous exercises. 432pp. 5⅜ x 8½. 0-486-66014-1

LINEAR ALGEBRA, Georgi E. Shilov. Determinants, linear spaces, matrix algebras, similar topics. For advanced undergraduates, graduates. Silverman translation. 387pp. 5⅜ x 8½. 0-486-63518-X

ELEMENTS OF REAL ANALYSIS, David A. Sprecher. Classic text covers fundamental concepts, real number system, point sets, functions of a real variable, Fourier series, much more. Over 500 exercises. 352pp. 5⅜ x 8½. 0-486-65385-4

SET THEORY AND LOGIC, Robert R. Stoll. Lucid introduction to unified theory of mathematical concepts. Set theory and logic seen as tools for conceptual understanding of real number system. 496pp. 5⅜ x 8¼. 0-486-63829-4

CATALOG OF DOVER BOOKS

TENSOR CALCULUS, J.L. Synge and A. Schild. Widely used introductory text covers spaces and tensors, basic operations in Riemannian space, non-Riemannian spaces, etc. 324pp. 5⅜ x 8¼.
0-486-63612-7

ORDINARY DIFFERENTIAL EQUATIONS, Morris Tenenbaum and Harry Pollard. Exhaustive survey of ordinary differential equations for undergraduates in mathematics, engineering, science. Thorough analysis of theorems. Diagrams. Bibliography. Index. 818pp. 5⅜ x 8½.
0-486-64940-7

INTEGRAL EQUATIONS, F. G. Tricomi. Authoritative, well-written treatment of extremely useful mathematical tool with wide applications. Volterra Equations, Fredholm Equations, much more. Advanced undergraduate to graduate level. Exercises. Bibliography. 238pp. 5⅜ x 8½.
0-486-64828-1

FOURIER SERIES, Georgi P. Tolstov. Translated by Richard A. Silverman. A valuable addition to the literature on the subject, moving clearly from subject to subject and theorem to theorem. 107 problems, answers. 336pp. 5⅜ x 8½. 0-486-63317-9

INTRODUCTION TO MATHEMATICAL THINKING, Friedrich Waismann. Examinations of arithmetic, geometry, and theory of integers; rational and natural numbers; complete induction; limit and point of accumulation; remarkable curves; complex and hypercomplex numbers, more. 1959 ed. 27 figures. xii+260pp. 5⅜ x 8½.
0-486-63317-9

POPULAR LECTURES ON MATHEMATICAL LOGIC, Hao Wang. Noted logician's lucid treatment of historical developments, set theory, model theory, recursion theory and constructivism, proof theory, more. 3 appendixes. Bibliography. 1981 edition. ix + 283pp. 5⅜ x 8½.
0-486-67632-3

CALCULUS OF VARIATIONS, Robert Weinstock. Basic introduction covering isoperimetric problems, theory of elasticity, quantum mechanics, electrostatics, etc. Exercises throughout. 326pp. 5⅜ x 8½.
0-486-63069-2

THE CONTINUUM: A CRITICAL EXAMINATION OF THE FOUNDATION OF ANALYSIS, Hermann Weyl. Classic of 20th-century foundational research deals with the conceptual problem posed by the continuum. 156pp. 5⅜ x 8½.
0-486-67982-9

CHALLENGING MATHEMATICAL PROBLEMS WITH ELEMENTARY SOLUTIONS, A. M. Yaglom and I. M. Yaglom. Over 170 challenging problems on probability theory, combinatorial analysis, points and lines, topology, convex polygons, many other topics. Solutions. Total of 445pp. 5⅜ x 8½. Two-vol. set.
Vol. I: 0-486-65536-9 Vol. II: 0-486-65537-7

Paperbound unless otherwise indicated. Available at your book dealer, online at **www.doverpublications.com**, or by writing to Dept. GI, Dover Publications, Inc., 31 East 2nd Street, Mineola, NY 11501. For current price information or for free catalogues (please indicate field of interest), write to Dover Publications or log on to **www.doverpublications.com** and see every Dover book in print. Dover publishes more than 500 books each year on science, elementary and advanced mathematics, biology, music, art, literary history, social sciences, and other areas.